T3-BLC-392

Laboratory Design Handbook

E. Crawley Cooper

CRC Press
Boca Raton   Ann Arbor   London   Tokyo

727.5
C772

Library of Congress Cataloging-in-Publication Data

Cooper, E. Crawley.
    Laboratory design handbook / E. Crawley Cooper.
      p. cm.
    Includes bibliographical references and index.
    1. Laboratories—Design and construction. I. Title.
TH4652.C66 1994
727'.5—dc20   93-44971
ISBN 0-8493-8996-8

This book contains information obtained from authentic and highly regarded sources. Reprinted material is quoted with permission, and sources are indicated. A wide variety of references are listed. Reasonable efforts have been made to publish reliable data and information, but the author and the publisher cannot assume responsibility for the validity of all materials or for the consequences of their use.

Neither this book nor any part may be reproduced or transmitted in any form or by any means, electronic or mechanical, including photocopying, microfilming, and recording, or by any information storage or retrieval system, without prior permission in writing from the publisher.

CRC Press, Inc.'s consent does not extend to copying for general distribution, for promotion, for creating new works, or for resale. Specific permission must be obtained in writing from CRC Press for such copying.

Direct all inquiries to CRC Press, Inc., 2000 Corporate Blvd., N. W., Boca Raton, Florida 33431.

© 1994 by CRC Press, Inc.

No claim to original U.S. Government works
International Standard Book Number 0–8493–8996–8
Library of Congress Card Number 93–44971
Printed in the United States of America 1 2 3 4 5 6 7 8 9 0
Printed on acid-free paper

Preface

This book is intended to assist those involved with creating new laboratories, remodeling existing laboratories or adapting an existing building to become a lab. Architects, engineers, scientists, facility managers, lab administrators, code officials, insurance underwriters, construction managers and builders may find it contains useful information. We have attempted to describe the process, motivation, constraints, challenges, opportunities, and specific design data related to the creation of a modern research laboratory facility. It is a reality that much of the information contained in this text will be completely outdated within a decade due to the rapid changes taking place in the sciences.

It is based on a large pool of experience in the development of new and renovated laboratory buildings for universities, teaching hospitals, pharmaceutical companies, start-up biotechnology companies and other types of industrial technology.

I am indebted to many people in gathering this information. Harry Orf, an organic chemist with the Massachusetts General Hospital in Boston and a Principal in Cambridge Laboratory Consultants, shared his knowledge with us when the MGH Lawrence E. Martin Labs in Charlestown, Massachusetts were under design. His partner at Cambridge Laboratory Consultants, Donald J. Ciappenelli, was kind enough to perform the technical review for this book. He made many valuable suggestions during that process. Robert Hsiung, a colleague at Jung/Brannen Associates, Inc., and designer of many outstanding laboratory facilities gave us some valuable insight into the design process. Richard G.

University Libraries
Carnegie Mellon University
Pittsburgh PA 15213-3890

Burnham of Dick Burnham Technical Sales and David Lupo of B & V Testing, Inc. contributed their valuable comments on the intricacies of lab equipment. Bruce MacRitchie of MacRitchie Associates, Inc. and Mike Zimmerman of Zimmerman Consulting contributed their engineering perspective. William Watson helped with advice related to the chapter on animal facilities. Ronald Prachniak, Corporate Projects Director at Massachusetts General Hospital, provided valuable insight on a wide range of research related issues.

The illustrations relating to chemical fume hoods and biological safety cabinets are reprinted by permission of the American Society of Heating, Refrigerating and Air Conditioning Engineers, Atlanta, Georgia, from the 1991 ASHRAE *Handbook–HVAC Applications*.

E. Crawley Cooper, AIA, earned his BS in Architecture at the University of Cincinnati and his Master of Architecture degree from the Massachusetts Institute of Technology. His current position is as Principal, Jung/Brannen Associates, Inc., in Boston, Massachusetts, and his previous experience was as Associate with Pietro Belluschi and Eduardo Catalano in Cambridge, MA; as Chief Designer with James Associates of Indianapolis, IN; and as Architect with Anderson Beckwith and Haible in Boston, MA. Mr. Cooper has served as Lecturer at the Massachusetts Institute of Technology, the Harvard Graduate School of Design, Purdue University, the University of Wisconsin, and at the Recombinant DNA Research Laboratory of the National Institutes of Health in Bethesda, MD.

In addition to directing numerous laboratory projects, Mr. Cooper has headed or been a member of groups such as school building committees, a long-range planning committee, a town planning board, a planning task force, and a re-zoning study committee. He has also acted as Consultant for Arthur D. Little, Inc., at the United Nations Conference Center in Vienna, Austria.

Table of Contents

Laboratory Design Handbook

Introduction

Firmness, commodity, delight: the three elements of architecture according to the great Roman architect and teacher, Vitruvius. This book has little to do with firmness or delight. It is mostly about commodity, or function, of the buildings where scientific research is conducted.

Modern research facilities provide usable space for laboratories, lab support areas, offices, and interactive spaces for formal and informal gatherings. The special equipment and environments required for research make these buildings extremely complex. A successful lab must also provide a safe and humane place for people to work. A well designed laboratory can be a significant tool for recruiting the best minds available. It can encourage the sharing of ideas in a culture that seeks the truth with an interdisciplinary team approach. And, laboratories are expensive facilities to build and operate.

The pace of discovery and the potential hazards of research have dictated that sophisticated mechanical and electrical systems and services are available to create pleasant, productive, and safe environments for scientific inquiry. It is not unusual for the building volume devoted to systems and services to exceed the usable, or served spaces. The relationship between the service volumes and served volumes in a laboratory facility is of paramount concern to the designer. However, a machine-centered design approach must be tempered by an understanding of how teams of researchers work and interact with each other and their environment.

The research scientist and the architect play complementary roles in the design process that results in a successful new laboratory. Generally, scientists are concerned with the micro scale. Studying a system's components in minute detail under a microscope is how they excel. Scientists are used to dealing with specifics; being precise is second nature to their culture. Things need to be quantified!

Architects, on the other hand, have been trained as generalists. Initially, they are inclined to look at the big picture. How will the project interact with the community, the available infrastucture, the environment? How will the laboratory users interact while performing their tasks? How can the proposed facility enhance and contribute to research? These questions are paramount to generalists. Issues need to be qualified and priorities established!

Before attempting to initiate a design concept, the scientists should prepare a mission statement about their work. What are their goals? How will they be achieved? Who will contribute in the achievements? What is the image for the facility? Identify the constraints. This exercise will help the research institution establish priorities and communicate among themselves and with their architect.

The architect should not have preconceptions about laboratory design. Often the best ideas for design concepts come from the building's users. The talented designer will listen to the mission statement for ways of creating opportunities out of challenges. Innovative concepts should be explored at both the micro and macro scale as a collaborative effort between the research scientist and the architect. By nature, many of the activities that occur in a laboratory setting are on the cutting edge of technology. The development of new and better processes and discoveries is the goal of research. Change, then, is inevitable over time. This establishes adaptability as a virtue for the lab facility.

Many research laboratories are developed around modular concepts that can provide maximum flexibility with a minimum of underutilized investment. The key ingredient is to provide appropriate structural volume arranged in a modular way that can accommodate a wide range of mechanical and electrical environmental systems.

Control of the environment is crucial to the success of scientific research. Over half of a "wet" lab facility construction costs are devoted to the mechanical and electrical systems. ("Wet" labs are defined as those labs with pure water and chemical fume hoods.) These systems contribute to the safety, reliability, efficiency, and productivity of the research work.

A generous vertical floor-to-floor dimension is essential for pharmaceutical or biological research laboratories in order to provide adequate space for the horizontal mechanical and electrical distribution systems. These systems can be placed above a suspended, accessible ceiling or within a dedicated interstitial service floor sandwiched between alternate laboratory floors. Either arrangement can be applied to a number of plan options. Obviously, the interstitial services floor concept is more costly, but it provides good long-term adaptability and a more efficient maintenance program. The Office of Management and Budget (OMB) under recent federal administrations would not permit federal funding for the interstitial floor concept because of the high initial cost premium for the larger building volume.

Wet labs consume staggering amounts of energy. In fact, they are energy "monsters."(One six-foot fume hood requires 1,100 to 1,200 cfm of 100% fresh, conditioned air 24 hours/day, year-round.) Strategies for reducing energy consumption include variable air volume (VAV) systems for the building and hoods, heat recovery techniques, such as plate exchangers and heat wheels, electronic direct digital automatic controls (DDC). Unfortunately, these relatively sophisticated measures only make a small dent in reducing operating costs and energy consumption.

Funding for research is expected to continue and grow over the next decade. In 1992, federal funding from the National Institutes of Health totaled 7.3 billion dollars. According to the Boston Redevelopment Authority, over the next ten years the medical research institutions in the Boston area alone will add an estimated 2.6 million square feet of research space and employ an estimated 3,100 to 5,500 new scientists and support staff.

1

"Hot" Areas of Research

Over the past several decades we have been experiencing the age of "microscience," research at the scale of the micron, one-millionth of a meter. Now, we are crossing the threshold from "microscience" to "nanoscience." The nanometer is 1,000 times smaller than the micron, or one-billionth of a meter. Current research is occurring at the molecular level without the aid of optical microscopes. This fascination with smallness and the quest to understand the basic building blocks of matter is changing the research laboratory in all fields of science, and it is integrating the sciences of chemistry and physics. Some of the major areas of research will occur in the following laboratory types.

Biomedical and Pharmaceutical

Drug research takes place on several different levels of experimentation. A drug's chemical reactions and structural modifications at the molecular level are the particular concerns of basic science. In the next stage, the pharmacological action of the drug on organisms is studied either at the cellular level (which includes viruses and bacteria) or in multicellular organisms (such as rodents and other mammals). Animal experimentation can be scaled to varying degrees of complexity, ranging from use of lower animals, such as worms, to more sophisticated studies with primates. Naturally, the more highly developed the experimental model, the more clinically applicable to humans are the expected results. With drugs that have selective pharmacological action, the organ affected, whether the liver, heart, or kidney, can be removed from the animal and studied as isolated tissue. Only after a drug has been thoroughly studied—not only for physiological effects but also toxicity—is human experimentation considered. The final step, clinical testing, involves administering the test drug to humans, according to the provisions of well designed guidelines provided by the Food and Drug Administration (FDA).

Each stage of drug research has its own significance, but the results do not necessarily carry over from one stage to the next. For example, the effects of the drug being studied on isolated rat liver in the test tube may be observed and measured. When the same drug is given to a living rat, however, the effects may not be similar. The rat liver in its normal anatomic setting is affected by multiple factors that may reinforce, distort, or otherwise alter the action of the drug being studed. Clinical testing is even more complicated. Humans are more complex than other animals, as well as chemically different in some respects. Futhermore, people have highly developed nervous systems, hence emotions and feelings, which affect them in elaborate and as yet poorly understood ways.

Biomedical and pharmaceutical research usually includes a vivaria, a special environment for holding animals or plants. The laboratories are wet labs with benches, sinks, pure water, chemical fume hoods, and standby power to protect the experiments.

Electronics and Computers

The backbone of modern technology is electronics. From sophisticated defense strategies such as "Star Wars," or SDI, to everyday consumer products like the personal computer, videocassette records and television, electronic and computer technologies are ubiquitous. A strong argument can be made that these technologies have improved our efficiency in information processing and communications to the point of causing massive personnel layoffs in businesses around the world. Nevertheless, the demand will remain strong to continue making improvements to these technologies.

Generally, electronic laboratories are considered "dry" labs, without much need for sinks and chemical fume hoods, although recently, experimentation is occuring in neural networks of electrochemical energy. Some scientists believe there may be the possibility of creating an electrochemical computer that would more closely mimic the brain's methods of processing data.

Optics and Lasers

Elimination of vibration is paramount for most facilities where optics are studied, especially using lasers. Light waves are measured in angstoms, a unit that is one ten-billionth of a meter. Stability is supreme in measuring in these units! Some types of laser systems require dedicated exhausts and chilled water cooling systems.

Magnets

Powerful electromagnets are used in physics, chemistry, and biology research. Controlled magnetic fields have proved to be a valuable tool in improving the technology of semiconductors, superconductors, and the imagery of living tissue. Electromagnets require powerful direct current rectifiers. They can also require tremendous cooling systems and a sophisticated shielding technology to contain the potentially harmful transmission of the high energy fields given off by the high voltage direct current. Some special magnets are operated under very low temperature conditions for superconductivity.

Materials

Materials science laboratories often have extreme temperature test chambers for both heat and cold. These labs usually require a high bay space with an overhead "H" crane for building large test mock-ups.

SUMMARY

Laboratory configurations can differ widely according to use and service requirements. Classification categories include discipline (chemistry, physics, biology), level of study (routine, teaching, research), equipment provisions (rigs, benches) and level of services (wet, dry).

Regardless of the classification, the modern research lab should not be so rigid as to accommodate just a single fixed use. Even the most short-sighted institutions are recognizing that rapid changes in technology demand adaptable, multi-use research facilities. New fields of study and changing state-of-the-art equipment require that adaptability be a prime concern in lab design.

One strategy for flexible planning is to group the wet waste drains that are gravity dependent, thereby preserving large areas with services only feeding from overhead. This will allow more freedom for future changes when adjusting to new technologies. Wherever possible, tables rather than fixed benches also make it easier to reconfigure space when necessary. The most important strategy for facility adaptability is to provide modular systems, where possible, that can expand incrementally, and to provide enough physical space initially to house that growth.

Planning a research facility cannot be a unilateral endeavor. Unlike other building types, ventilation and exhaust systems, the lungs of the beast, play a primary role in determining how the various spaces are organized. Cost trade-offs must be considered at every turn. Occupant safety is a major concern, as with any building type, but with the presence of potentially hazardous materials and processes, safety must be considered with every decision. Zoning issues, institutional regulations, and code constraints need to be thoroughly analyzed initially. Above all, good communications among all participants, especially the scientists, administrators, engineers, safety officer, cost estimators and architect, are required.

2

The Planning Process

Compared to most building types, research labs are expensive. The high unit costs of these facilities reflect architectural features such as casework and special finishes, as well as the environmental requirements for special equipment, HVAC, power, and other laboratory services. In addition, the dynamic nature of research demands flexibility in facility design responding to continuing change in occupancy and use. Consequently, planning for research laboratories requires a careful analysis of both space requirements and specific services needs for each space.

The planning process is designed to encourage the users/occupants of the facility to determine the design requirements. Architects must provide an organized process and technical resources to assist the research teams in converting their requirements into an initial definition of building design criteria from which it is possible to derive an order of magnitude estimate of probable cost. The user's needs are evaluated to identify duplication, suggest alternatives, and advise on the cost/benefit trade-offs of significant elements in the project criteria. If possible, various initial baseline design concepts are tested and compared.

Often the architect is a mediator between conflicting agendas of administration and staff, while working within budget limitations. This setting of priorities can be a real balancing act. Much time is spent soliciting ideas from the researchers. Some of the best design ideas are often generated by users. Functional relationships are given more attention than aesthetic aspects of design at these early stages. While the architect must serve as a good listener, there are times when his/her role is to challenge the clients to stretch their thinking. Sometimes the user does not fully comprehend the design issues and opportunities. Open communications are essential to develop a team approach that will result in a product that everyone can "own."

Objectives

A clear definition of the researcher's goals and objectives is the fundamental first step in planning a facility. Ideally, the researchers will prepare a written mission statement. These goals need to be documented and "owned" by the entire design team, including the administration, managers, and researchers. These objectives may include the desire to enhance group interaction, the need for the project to express a positive image to the community, and/or enhancement of the user's recruiting efforts.

Being a good listener is essential in communication. The nonscientists need to understand the language of buzzwords found in the research community. Accordingly, we have included a glossary of terms in the Appendix to help the uninitiated in their listening process.

Programming

Work space efficiency, economy, and productivity are directly impacted by the thoroughness of these crucial planning steps. Administrative and operational factors must be understood, analyzed, and clarified, including the research philosophy (individual or team oriented), security, data management procedures, use of robotics, extent of shared services such as food service, library, conference spaces, storage, purchasing, glass washing, materials handling, and radiation applications.

Data Collection

Intensive onsite interviews and investigations are used to develop a specific set of programming requirements for the project where specific space requirements are expressed in terms of variations from the baseline design criteria. Customized questionnaire forms are developed as a tool for this task.

Space requirements are identified, cataloged and analyzed using several methods:

- consensus-based, utilizing interview techniques
- productivity-based, utilizing the institution's strategic plan
- measurement against industry standards

Analysis

The programming data are evaluated and converted into a comprehensive facility plan including tabulation of space requirements, design criteria, adjacency and relational matrices, bubble diagrams, and preliminary probable project cost estimates and operating cost pro formas.

Functional relationships or adjacencies are determined by such factors as safety, work flow, shared services, and staff interaction. Key components to be analyzed and documented include:

- lab to lab
- lab to shared space
- lab to shared equipment
- lab to offices
- material deliveries and waste disposal

Key laboratory components to be analyzed include:

- lab workstations
- staff offices
- support space
- mechanical service space
- corridor points of entry and egress

- daylight contributions
- critical environmental concerns, i.e., vibration, air pressure, humidity

Before an informed decision can be reached on facility design involving hazardous production materials (HPM), programming must include product definition, process, and process equipment requirements and the development of a strategy for the delivery, storage, and use of HPM as defined in the Uniform Building Code and the Uniform Fire Code. The estimated quantities of HPM (liquids, solids, and gases) to be stored and used must be determined as precisely as possible. This will determine code requirements for facility occupancy, size, and egress patterns.

Codes and Standards

A thorough search of governing zoning restrictions, environmental regulations, building codes, underwriter's requirements, institutional and industry guidelines and standards is necessary to refine the program's needs and set design constraints. Generally, the permitting and project approvals process for a research laboratory is initiated during the programming phase because of the time required by the reviewing agencies.

Hazardous materials are tightly regulated as to how they are transported, stored, dispensed, used, and disposed. Some requirements regarding these materials include:

- explosion venting for flammable and pyrophoric materials
- smoke detection and venting
- treatment systems (diluting, absorbing)
- fire resistant construction
- ventilation and exhaust
- alarm systems
- electrical systems classification
- segregation of incompatible materials
- fire suppression systems
- exterior separation distances

- floor plate area
- access signage
- egress pathways

It may be important to involve the local building official, fire marshal, air quality management agency, and other appropriate agencies during the early programming phase to avoid serious problems later.

Engineering Systems

Affecting the facilities function, flexibility, and economy are the amount and types of mechanical and electrical services needed and how they are provided. Multiple and complex systems are required for a modern laboratory. The functional and spatial integration of these various systems to the structural framing system and personnel circulation network is crucial to a successful laboratory design. Baseline building systems alternatives are considered with their associated space needs and probable cost estimates. Special, area-specific, systems are also defined, compared, and analyzed.

Initial Cost Projections

As programming requirements begin to solidify, a probable cost projection is developed using factored unit costs based on similar completed projects. It is important to adjust the figures for escalation and location factors.

Feedback

Feedback from the users and management is critical to the planning process. The results of the initial analyses and cost projections are reviewed with the researchers and managers who provided information, to confirm interpretations and to advise them of recommended changes or variations in criteria. Here, the effort focuses on the process of evaluating and recommending trade-offs to meet budgetary restrictions.

Final Program

The information gathered during the feedback process is incorporated into the final program. The final program will include space characteristics definition for each laboratory and office group, adjacency and relational data presented as matrices and bubble diagrams, design criteria covering codes, utility requirements, environmental issues, and a project probable cost estimate. A written program of facility requirements should be the result of interactive meetings between the various user groups, administrators, and the design team. Such a document allows all participants to review and understand the design issues prior to trying to reach a particular concept or solution.

However, by setting forth trial concept options during the programming phase, the architect can stimulate thinking, develop lists of advantages and disadvantages, and begin to make cost comparisons for initial and operational costs. These early test concepts are an important element of the programming task. They can influence site selection, point out potential building code constraints, and highlight the need for obtaining input from special consultants where in-house expertise is lacking.

These steps serve as the basis for a specific work plan developed to suit the requirements of a particular project.

Work Plan

A typical programming work plan is outlined below. While the timing and specifics may change from project to project, the basic steps will be common to all projects.

Objectives

Week 1: During the first week a site visit will be made by the planning team for the purpose of gathering data on existing conditions and establishing project policies and goals.

- Meet with laboratory and facility management to establish project policies, goals, objectives, opportunities, and budgets. Determine the image, aspirations, and overall objectives desired for the research staff and its facilities.
- Research and product functions
- Organization and staff policies
- Implementation constraints and priorities
- Collect applicable physical plant information on existing research facilities and structures which may be involved with the new facilities.
- As-built construction drawings for original construction and major renovations
- Topographical and geotechnical data for the site
- Utility data, including capacities and rate structure
- Facility operating and maintenance data, including staff, energy and utilities consumption and costs
- Document existing physical plant information and space use and characteristics.
- Inspect existing mechanical and electrical systems and document existing conditions and problems.
- Identify all spaces by use and department, record area, general equipment list and layout.
- Establish present environmental conditions, utility needs and special requirements, including hazardous materials and operations.
- Determine regulatory agencies with jurisdiction and identify applicable codes and regulations. Obtain corporate standards that are applicable.
  - Local: zoning, fire, building, environmental
  - State: fire, industrial health and safety, building, handicapped
  - Environmental
  - Federal: OSHA, NIOSH, EPA, NRC, FDA
  - Industry guidelines: NIH, GLP, ASHRAE, SCMACMA
  - Insurance: Underwriter's, NFPA

- Obtain master departmental staffing (including job descriptions) and organization charts.

Weeks 2 & 3: Review and evaluate existing conditions data and develop master computerized database of existing facilities.

- Evaluate facility conditions; identify substandard laboratory working space conditions, safety problem areas, and special needs.
- Develop baseline laboratory and office chart characteristics and unit cost data.

Programming

Week 4: Second site visit by planning team for purpose of establishing detailed program requirements for new facilities.

- Review existing space conditions with uses to determine inadequacies and shortcomings that need to be addressed in the new facilities.
- Identify major new equipment and functions that need to be accommodated in the new facilities.
- Planning team to identify:
  - Space functions, size, quantity, relationships to other spaces and/or functions.
  - Proposed equipment and furnishings schedule with determination of utility, space, and systems requirements.
  - Proposed design and environmental criteria—structure, power, communications, HVAC, lighting, hazardous and special materials handling and supply, vibration and acoustic control.

Weeks 5-7: Review and evaluate program data collected in the programming interviews.

- Develop master computerized database for new facilities program by revising database developed for existing facilities.

- Develop and document space characteristic definitions for each laboratory and office group.
- Analyze program data to determine system design criteria and appropriate baseline space characteristic changes. Update base unit cost data.
- Summarize complete utility system capacity requirements. Identify major users of utilities to isolate cost-driving factors.
- Develop adjacency requirements based on both organizational needs and environmental conditions. Develop bubble diagrams and stacking diagrams reflecting organizational options.
- Develop facility requirements statement with related costs identified. Provide a comparison with program budget objectives. Identify cost-driver elements.

Feedback

Weeks 8 & 9: Third site visit by planning team for the purpose of reviewing the facility program with the users and management. The objective of this review is to confirm the interpretation of the users' requirements and to present planning options for evaluation by the users and management.

Final Program

Weeks 10 & 11: Incorporate data developed during the feedback sessions into the final planning report, prepare final documentation and submit to client.

3

Design Options & Opportunities

Planning Module

The planning module for a laboratory is usually determined by the bench spacing, and the bench spacing is determined by the anthropomorphic dimensions of the lab worker and his/her work space (Figure 3.1). The minimum work space per person is about 5 feet long by 51 inches deep. This depth is made up of a 27 to 30 inch work surface plus a 21 to 24 inch space for the seated or standing worker. In addition, a passage space is necessary behind the worker of about 18 to 24 inches, or more if carts must fit through.

A lab module width which allows two parallel rows of benches with a center gangway providing room to pass between back to back workers usually falls between 10 and 11 and a half feet. If this passage width is too great, the lab workers will be tempted to place tables or equipment between the benches, which of course would create an unsafe condition. Therefore, be skeptical of the lab module that exceeds 11 feet, 6 inches.

Module length is determined by safety considerations. Three workers along a lab bench are considered the maximum before a cross aisle is provided. Since benches are often placed in a series of peninsulas with access from one end only, the module length usually falls in the range of 16 to 27 feet. Therefore, a bench space module for two researchers can be accommodated in as little as 160 ft^2 (10′ × 16′). However, they also require adjacent support space. The minimum gross building area for each lab worker is 325 ft^2, according

Figure 3.1 Generic Bench Space. The basic laboratory planning module width is based on (1) work surface depth, (2) worker space, and (3) access space between back-to-back workers. Photo shows peninsula with bench space on both sides of shelving mounted above the benchtops.

to NIH funding standards. However, many labs exceed that minimum standard.

Support Zone

Lab support activities occur more appropriately away from the bench area, but close by. Generally, these support spaces are for special equipment or for work tasks that require unique environments such as fluorescent microscopy, tissue culture, or dark rooms. Noisy and high heat generating equipment such as ultra low temperature freezers and centrifuges are better located slightly remote from the bench work area. It is not unusual for this lab support space to equal or exceed the area of the generic lab benches.

Corridors

A clear width of 6 to 8 feet is common for laboratory corridors where tanks, equipment, and carts must traverse. However, where a separate dedicated service corridor is provided, the personnel passageway can be reduced to 5 feet. Such a narrow width is probably a good idea, because it will inhibit the potential for placing refrigerators or other storage items in the corridor. It is recommended that the secondary lateral corridors, or virtual corridors within the lab at the end of the peninsula benches, contain a 3 foot wide equipment zone on the opposite side of the passage from the bench ends. This provides a good location for shelving, work tables, refrigerators, etc. for easy access by the bench worker.

Basic Designs

There are several basic design concepts for the modern research laboratory; to wit:

- A central double-loaded service/shaft spine flanked on both sides by interior (windowless) laboratory space. Beyond the lab zone is a double-loaded personnel corridor that feeds the labs and perimeter offices with windows. This is commonly considered today's ideal model for lab design, since it ranks high in all aspects of functionality, safety, and cost.

- Where site constraints limit potential building depth, a single-loaded service/shaft version of the aforementioned model is a realistic option. Sometimes, this concept is called the "exostitial space" design concept, since the services are along the exterior wall of the laboratory zone.

- Alternating floors dedicated to the systems services between programmed laboratory floors is known as the "interstitial space" design concept. In the previous decade it was considered by many to be the ideal model for organizing laboratory space. It has become somewhat unpopular because of the poor net to gross floor usage efficiency and corresponding high initial cost. During the mid-1980s, the Office of Management and Budget (OMB) of the executive branch of the federal government refused to provide capital funding for the "interstitial space" because of the inherently higher initial construction costs.

- Distributed mechanical equipment rooms and shafts at each floor is a kind of single-loaded interstitial concept that minimizes the central plant equipment in favor of mixing these spaces among each of the laboratory floors. Usually, this requires fresh air intakes at the perimeter of each floor.

Any of the above design concepts usually has some central systems equipment located at the base and top of the building. Inevitably, the fume hood exhausts should exit at the roof. Obviously, the fresh air intakes should be remote from the contaminated exhaust effluent.

Option A

A long, clear structural span between the exterior wall and a 14 to 16 foot wide continuous interior service corridor/shaft, resulting in a basic building width of about 130 feet (Figure 3.2). The clear strip can be developed in a number of arrangements including typical lab bench arrangement,

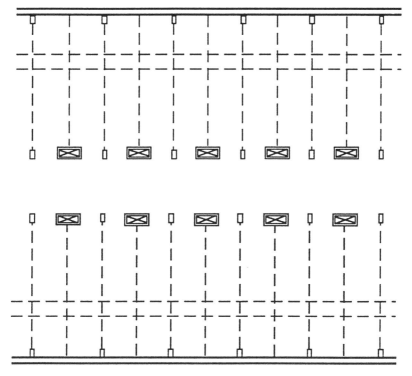

Figure 3.2.   Partial Floor Plan—Option A.

along the exterior with adjacent specialized lab support spaces or a 12 to 16 foot deep office strip at the windows, a 5 to 8 foot clear corridor, and 32 to 38 foot deep laboratories. Generally, offices will be laid out on a five or five and a half foot module, while labs will be developed on a 10 to 11 foot module. A 16 foot floor to floor height will allow 10 foot clear lab ceilings and adequate ceiling space within the deep trusses of the long span for mechanical services.

Pure and nonpotable water, power, compressed air, natural gas, oxygen, nitrogen, vacuum, other gases, and drains are distributed through the service corridor/shaft and run out to the lab spaces above the ceiling, or in lab casework as the individual lab layout requires.

HVAC systems are developed on a modular basis using

individual fan coil, constant or variable volume air handling units for lab and office cooling and heating. Units are located on the floor served adjacent to the service corridors above the ceiling. Hot water and chilled water are distributed along the continuous service corridor/shaft.

The advantages of this option are:

- Most alteration work is done in the service corridor/shaft, which minimizes disruption to ongoing research.
- Some overhead modifications can be achieved with temporary staging spanning between the building trusses, thereby creating less disruption to the building users.

The disadvantages are:

- Service corridor could inhibit potential for user interaction from one side of the floor to the other.
- Structural deflections inherent with long spans may require localized vibration isolators or for certain types of instrumentation or laser work.

Option A-1

This option is similar to Option A, with the addition of a series of small shafts along the exterior wall containing the risers for the systems serving the 14 foot deep perimeter zone (Figure 3.3). This design has a slightly higher initial cost but it provides additional flexibility by allowing more shaftspace distribution.

Option B:

Utilizes the same plan but introduces another row of columns to reduce the structural span (Figure 3.4). A central service corridor is flanked on either side with service shafts. A series of smaller shafts could enclose system risers at the exterior wall.

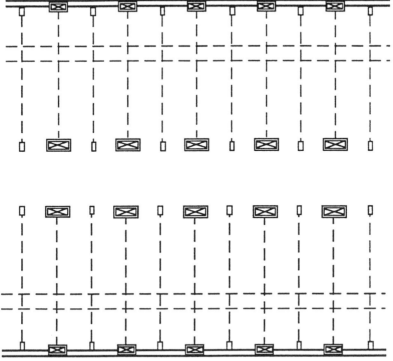

Figure 3.3.   Partial Floor Plan—Option A-1.

The advantages of this option are:

- Shorter structural spans are more economical for reducing floor-to-floor height and for controlling vibration and deflection, a necessary consideration for research involving lasers and electron microscopes.
- Good potential for open laboratory plan and for occupant interaction.

The disadvantages are:

- Additional structural columns are somewhat inhibiting for future planning changes.
- Limited interaction between occupants on opposite sides of service corridor.

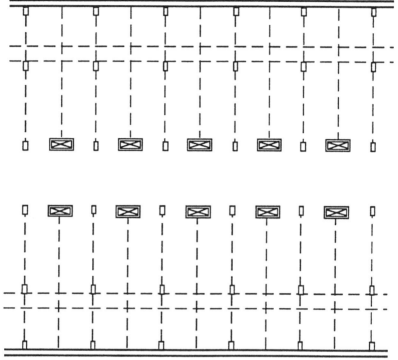

Figure 3.4. Partial Floor Plan—Option B.

Option C

This has the vertical shafts spaced only along the exterior building wall (Figure 3.5). It also incorporates long structural spans with deep open trusses to provide uninterrupted laboratory space.

The advantages of this option are:

- Unobstructed interior permits planning flexibility and occupant interaction.
- Some overhead modifications can be achieved with temporary staging spanning between the building trusses, creating less disruption to the building users.

The disadvantages are:

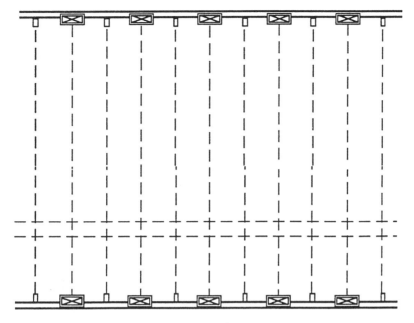

Figure 3.5. Partial Floor Plan—Option C.

- Waste and water services may require freeze protection.
- Changes to risers within the shafts will be very expensive.
- Window treatment and views are inhibited.
- Structural deflections inherent with long spans may require localized vibration isolators for certain types of instrumentation or laser work.

Commonalities

Central building systems equipment, i.e., make-up air units, substations, standby or backup generators, compressed air supplies, vacuum pumps, pure water source, etc., are located in the ground and second floor mechanical rooms and distributed vertically through the service shafts. Laboratory service systems are developed in a modular basis so that capacity can be added as necessary. Central chilled water and steam or hot water sources at the ground floor will provide all heating and cooling energy for each building (Figure 3.6).

Figure 3.6.   Equipment zone along secondary passage at the end of bench peninsulas. Note ice machine in foreground, just in front of emergency shower and eyewash.

Pharmaceutical and biotechnical laboratory buildings require large lungs. Fume hood exhausts and clean, sterile spaces require large quantities of very controlled, conditioned air. Maintenance of pressure differentials between certain spaces becomes crucial for contamination prevention and safety. Fresh air intakes must be remote from exhaust discharges. Thus, all design options introduce fresh air into the building at the ground or second floor. Fume hood exhausts rise through the shafts to a fan gallery on the roof where exhaust fans and emissions control equipment can be located.

Laboratory building efficiencies can vary widely when compared to other building types. The use of more hazardous materials generally results in less efficiency (ratio of net to gross areas). Labs requiring ultra-clean environments will be less efficient than those without clean rooms. A comparison of some 20 recent medical research facilities across the country revealed the average net usable area amounted to only 64% of the total building gross area.

Provisions for the Disabled

The American with Disabilities Act (ADA) prohibits discrimination against a disabled individual in either employment, availability of goods and services, transportation, communications, or in access to any place of public accommodation. This is a law, not a building code regulation. It is applicable to every facility in the United States, except residential units, religious organizations, or private clubs. According to Congress, the purpose of the law is: "to provide a clear and comprehensive national mandate to end discrimination against individuals with disabilities and to bring persons with disabilities into the economic and social mainstream of American life."

All emergency devices and equipment should be accessible to disabled persons. At least one chemical fume hood on each floor needs to have appropriate knee space below the sash for an operator seated in a wheelchair. At least one bench workstation should have a lower countertop (32") and knee space. All of the other ADA compliance features should be met, such as 36 inch egress pathways, 60 inch turning radii for small rooms, including environmental and microscope rooms.

Adaptability

What is the meaning of adaptability in a laboratory building?

- Creation of flexible labs, special-use labs, and office space with minimal modifications to the building systems.
- Simplified and straightforward base utility systems which allow for staging and sequencing future changes with minimal impact to the continuing lab functions.
- Adequate space for future equipment and systems expansion within the overall building enclosure.
- Where possible, modular equipment used, with generous network distribution system capacities.

Renovation Challenges

A key strategy for retrofit construction is the development of a staging or sequencing philosophy of providing temporary or permanent space for ongoing functions so that the necessary dismantling work and new construction can occur in a programmed and logical way. Such a strategy might be enhanced by the use of entirely new space for systems distribution, thereby permitting an orderly, staged construction process.

One of the most successful methods of modernizing an older building is the use of concepts known as interstitial and/or exostitial mechanical/electrical services space. Both of these concepts will require increased floor areas for the systems distribution networks. The interstitial space concept requires the use of existing usable space. Interstitial space is commonly seen as using a floor or mezzanine between other floors to locate new equipment and networks. Using the interstitial space concept means that existing space must be sacrificed to achieve the ultimate objectives.

Exostitial space is the introduction of additional building area at the perimeter or on top of existing space. Although the cost of providing this service space represents a capital expense for nonprogrammed activities, the concept allows significant flexibility with respect to the utility corridors.

Either concept permits future space changes to be a "plug-in" situation mitigating the expense of repeated system modifications from distant sources. Both of these concepts (interstitial/exostitial) basically achieve the objective of simplifying the next generation of space changes.

The following objectives need to be considered in any laboratory retrofit project:

- initial construction cost
- total disruption time
- impact on cash flow
- code and ADA compliance
- future flexibility to modify programs
- business interruption
- need for temporary "swing" space
- ability to provide dedicated service/communications corridors
- ease of future maintenance
- space organization efficiency
- future retrofit costs/time
- ability to maintain a high level of safety.

4

Laboratory Safety

Research laboratories can be dangerous places. To engage in research is to extend the edges of knowledge, to experiment. Hazards often accompany these investigative tasks. These hazards can include ignitability (fire, explosions), corrosivity (acids), reactivity (acid/base or polymerization), toxicity (ingested or absorbed contaminants) and, of course, electrocution. Radioactive materials, carcinogens, infectious diseases, viruses, and toxic gases are commonly found in the biology or medical laboratory setting.

Most laboratories are defined as hazardous waste generators by the latest issue of the Resource Conservation and Recovery Act (RCRA) of the United States, which means that the lab waste products (solids, liquids, and gases) must be stored, transported, and disposed of in a carefully controlled manner. RCRA regulations and procedures are a matter of law. Hazardous materials delivered and stored in laboratories and the waste products from research activities should be of tremendous safety concern to building occupants and the general public. Wherever possible, quantities of these materials should be kept to practical minimums in the laboratory facility and safely stored in small, properly labeled containers.

Elements of Health and Safety

Training	Exposure
Laboratory Safety	Radiation Safety
Noise	Ergonomics
Emergency Response	Respiratory Protection

Engineering Controls	Air Quality
Hazardous Wastes	Chemical Controls

Hazards

Toxic	Flammable, Pyrophoric
Corrosive	Explosive
Radioactive	Reactive (with water)
Sensitization	Infectious
Irritation	Communicable
Teratogenic	Carcinogenic
Extreme Temperatures	Mutagenic

Emergency Response Issues

Life Safety: Reduce injury or exposure to hazardous materials.

Environmental Protection: Reduce release of hazardous materials via air, water, or solid waste dispersion.

Property Damage: Reduce physical damage from fire, explosion, or hazardous chemical release.

Liability: Reduce risk of fines or laws due to improper or negligent operations.

Biosafety Levels

Biosafety levels consist of laboratory practices, safety equipment, and laboratory facilities. Microbiological and biomedical laboratories are categorized by the National Institutes of Health (NIH) and the Center for Disease Control (CDC) into four possible biosafety levels. These levels require four different combinations of operational techniques, safety equipment, and facilities design appropriate for the hazards posed by the laboratory activity. The following are requirements common to all biosafety levels:

- Surfaces must be easily cleaned.
- Horizontal work surfaces must be impervious to water, resistant to acids, bases, solvents, and heat.

- Furnishings must be sturdy.
- Spaces between benches, cabinets, and equipment must be accessible for cleaning.
- A hand-washing sink must be available in each lab.
- Any operable exterior windows must have insect screens.

The levels and their characteristics are as follows:

Biosafety Level 1 (BL-1):

"...appropriate for undergraduate and secondary educational training and teaching laboratories and for other facilities in which work is done with defined and characterized strains of viable micro-organisms not known to cause disease in healthy adult humans."

No special containment safety equipment is required. Work is generally performed at open bench tops. No recombinant DNA activities are permitted in BL-1 facilities.

Biosafety Level 2 (BL-2):

"...applicable to clinical, diagnostic, teaching and other facilities in which work is done with the broad spectrum of indigenous moderate-risk agents present in the community and associated with human disease of varying severity.... Primary hazards to personnel working with these agents may include accidental auto-inoculation, ingestion, and skin or mucous membrane exposure to infectious materials. Procedures with high aerosol potential that may increase the risk of exposure of personnel must be conducted in primary containment equipment or devices."

In addition to the requirements listed for BL-1, access to a BL-2 facility is limited to those permitted by the laboratory director. Warning signs incorporating the universal biohazard symbol must be posted at the lab entrance along with the lab director's or other responsible person's name and telephone number, and any special requirements for entering. Laboratory coats or smocks should be worn while in the lab and removed when leaving the lab or covered with a clean coat not used in the laboratory. Wastes should be appropriately decontaminated prior to disposal. Recombinant DNA research is permitted and biological safety cabinets are required for some procedures.

Biosafety Level 3 (BL-3):

"...applicable to clinical, diagnostic, teaching, research, or production facilities in which work is done with indigenous or exotic agents where the potential for infection by aerosols is real and the disease may have serious or lethal consequences. Auto-inoculation and ingestion also represent primary hazards to personnel working with these agents."

The BL-3 facility may be a single laboratory suite or a complex of modules within a building, or an entire building. It should be separated by a controlled access zone from areas open to the public and to other laboratory personnel who do not work within the BL-3 area. The BL-3 lab has special features that make it possible for lab workers to handle moderately hazardous materials without endangering themselves, other resident personnel, the community, or the environment.

Access to the facility must be through a lockable sally port vestibule arrangement with self-closing doors interlocked so that only one door can be open at a time. Neither door should be opened when experiments are in progress. The lab must be under negative pressure with respect to the surrounding areas and the outside. There must be a visible indication of directional air flow at the entrance sally port doors. (An economical, maintenance-free indicator is to drill a 1/4 inch diameter hole in the doors and have telltale yarn show any air movement direction like a wind sock.) It is recommended to have an automatic alarm sound in the event the negative pressure is compromised. Besides the ventilation alarm, it is good practice to interlock the supply and exhaust fans so that should the exhaust fan fail, the supply fan automatically shuts down to prevent an accidental positive pressure condition. In any event, the exhaust fan should have a backup power source.

Conditioned air is supplied through nonaspirating (laminar flow) type ceiling diffusers in a manner that creates unidirectional air flow from spaces of lower contamination potential to spaces of higher contamination potential without disrupting the air flow at the face of the biosafety cabinets. HEPA filtered exhaust ducts from Class I or II biosafety cabinets may be discharged into the lab provided the filters are tested and certified every year. It is probably safer to exhaust this HEPA filtered air directly outside. The exhaust air from BL-3 facilities should be discharged to the outdoors clear of occupied buildings and their supply air intakes. This is usually accomplished by locating exhaust stacks on the roof and exhausting upward at a velocity of at least 3,000 FPM. The balance of the air not exhausted through the biosafety cabinets may exhaust to the exterior without HEPA filtering but for greater safety it too should be HEPA filtered. Housings for filters should be bag-in, bag-out. Both the supply and exhaust ducts need to have accessible isolation dampers outside the facility to permit space closure for decontamination, generally with formaldehyde gas.

The finishes for walls, floors, and ceilings should be resistant to liquid penetration and be readily cleanable. Floors should be seamless, and if bases are not coved, they should be sealed to prevent leakage of liquids to adjacent areas. Suggested ceiling materials are gypsum board or plaster with epoxy paint finish.

Mechanical and electrical distribution networks above the ceiling should be kept to a minimum in BL-3 labs, and valves, dampers, etc. should be accessible from above the accessible ceilings of adjacent spaces. If windows are provided, they must be sealed shut. The openings in walls, floors, and ceilings through which utility services and air ducts penetrate should be tightly sealed with sanitary type silicone sealant to provide airtight enclosure necessary for decontamination procedures.

A foot, elbow, or automatically operated hand washing sink should be located near the exit door of each primary

BL-3 module. An autoclave must be located within the BL-3 facility. With appropriate procedural controls, it is possible to place the autoclave in the air-lock vestibule. If a pass-through autoclave is installed, most of the mechanical parts should be outside the lab for heat reduction and ease of maintenance. An exhaust register should be placed over the autoclave door to minimize humidity buildup within the lab.

Stainless steel bench surfaces are preferable. An emergency eyewash and shower should be readily available to the lab personnel, as well as fire extinguishers. The building fire alarm system must be audible to the occupants of the BL-3 laboratory. Special protective clothing (solid front or wraparound gowns, scrub suits, coveralls) are worn inside the facility only, and decontaminated prior to laundering or disposal. Disposable shoe covers are to be worn in the lab and removed upon leaving. Gloves are to be worn when handling infected materials and double gloves while working in the biosafety cabinet, with the outer pair being removed before withdrawing hands from the cabinet. Wall hooks near the lab entrance are necessary for the protective clothing. Some BL-3 labs have a shower, unisex toilet room and changing area as part of the sally port entrance, and under certain conditions workers are required to shower in and shower out.

Vacuum lines serving the BL-3 lab must be protected with HEPA filters and liquid disinfectant traps. Molded surgical masks or respirators are worn in rooms containing infected animals or plants. Experiments are conducted only within the containment cabinets and the work surfaces are decontaminated at least daily, and after any spill. Any recombinant DNA research involving human pathogens must be conducted in a BL-3 laboratory.

Biosafety Level 4 (BL-4):

"...applicable to work with dangerous and exotic agents which pose a high individual risk of life-threatening disease.

All manipulations of potentially infectious diagnostic materials, isolates, and naturally or experimentally infected animals pose a high risk of exposure and infection to laboratory personnel."

The BL-4 facility should be in a separate building outside of an urban area, or at least in a completely isolated zone of a building. Access must be very restricted, and a logbook signed by all entering and leaving personnel. Personnel should probably have a complete semiannual physical examination. Lab workers enter a BL-4 laboratory through a clothing change room where all street clothing is removed, including underwear and shoes. Special lab clothing is worn within the lab where only glove box procedures are performed. Otherwise, one-piece pressurized body suits are to be worn similar to space-walking astronauts. The use of these suits permits larger scale experiments and allows more flexibility in the size of the test equipment. Nevertheless, the BL-4 laboratory must be tightly sealed with sophisticated life support systems for the occupants.

The BL-4 lab at the Center for Disease Control in Atlanta has a super-safe breathing system for the pressurized suits. Two oil-free, water ring compressors generate air that passes through a dryer that humidifies and cools the air. A large tank holds a prescribed level of air in case there is a period when the compressors are malfunctioning. Ceiling-hung coiled air hoses feed the air to the "space" suits. The tanks hold a supply of air that would last long enough for lab evacuation in the unlikely event of a total power failure, including the backup generators (Figure 4.1).

The interior finish surfaces become the containment barrier, because they are the easiest to inspect, test, and repair. To test the containment seal, a negative air pressure is created after applying a liquid detergent to all surfaces. Any leak causes a bubble to form in the detergent, so the point can be marked and re-sealed.

For every usable square foot in a BL-4 facility there are at least two square feet of mechanical and electrical equipment support space. The negative pressure must be maintained at all times within the BL-4 laboratory with the exhaust air

Figure 4.1.   Gas tanks with wall hitch to prevent toppling.

HEPA filtered and incinerated. Manometers are used to monitor the pressure differentials and sound the alarm if they are too low; and of course, the supply and exhaust fans should be interlocked so that there is no possibility of creating a positive pressure condition should the exhaust fan malfunction. The only time a BL-4 laboratory is not under negative pressure is during a period of decontamination using a lethal gas.

No eating, drinking, smoking, storage of food, or application of cosmetics is permitted in BL-1, BL-2, BL-3, and BL-4 labs.

Animal Safety Levels

There are also four biosafety levels associated with activities involving infectious disease with experimental mammals (ABL-1, 2, 3, and 4) which also require increasing levels of protection to personnel and the environment.

Prime Responsibility

Each biosafety level has extensive recommendations with precise criteria dealing with lab practices, safety equipment, and laboratory design details. The lab director is directly and primarily responsible for laboratory safety, but the laboratory design architect also should be thoroughly familiar with the NIH and CDC published standards and recommendations. These are available from the U.S. Department of Health and Human Services.

Occasionally, a BL-2 laboratory will contain a space or suite of spaces that are classified for BL-3 research work with toxic materials which can cause disease if the organism enters the body. The purpose of containment is to reduce exposure of these bacteria or viruses to the lab occupants and to prevent the release of the materials into the environment. Containment is achieved through a combination of the use of biosafety hoods which act as the primary containment, and strict operational procedures. The BL-3 lab itself provides secondary containment through physical enclosure with air locks and controlled negative pressure in relationship with surrounding spaces. The pressure differentials between vestibules, corridors, air locks, and the laboratory facility, and across the HEPA filters should be continuously monitored and alarmed. The designer's challenge is to achieve a controlled pressure relationship in these spaces for all possible modes of usage and occupancy.

It is required to use HEPA filtration and 100% outside air

for a BL-3 facility. It is very important to provide tight seals at all penetrations of the containment boundary, not only for normal operation, but also for decontamination (fumigation) when poisonous formaldehyde gas is used to kill all living organisms. Plumbing connections from a BL-3 lab should be routed to a kill tank within the containment barrier; otherwise, materials are autoclaved and sterilized within the BL-3 lab.

Guidelines for Toxic Materials Handling

The Uniform Building Code (UBC) for Group H, Division 6, laboratory occupancy requires that fire exit corridors be separated from other laboratory areas by at least a one-hour fire resistive barrier. It also prohibits the transport of hazardous production materials (HPM) in exit corridors. The dedicated service corridors must be mechanically ventilated at not less than 6 air changes per hour. The UBC requires additional safety features, including limits on HPM container sizes, use of approved transport carts to carry gas cylinders to the labs, and installation of emergency alarms at 150 foot intervals. HPM cannot be stored or dispersed within the exit passageway proper.

The H-6 requirements are the UBC embodiment of California's "Green Book," that since 1981 has set design standards for electronics and semiconductor facilities with onsite hazardous materials. Adopted January 1, 1985, H-6 governs all facilities using highly toxic chemicals or gases found in Table 9A, line 18 of the Uniform Building Code. These include: arsine, phosphene, silane, hydrogen, hydrofluoric acid, and others.

HPM storage room for dispensing Class I or Class II flammable liquids or gases cannot exceed 1,000 ft^2. Such a room must have a two-hour fire resistant enclosure when the room area exceeds 300 ft^2. A one-hour fire separation is required for HPM storage rooms less than 300 ft^2 in area.

Meeting H-6 code requirements can significantly increase the net to gross area efficiency ratio of a building, thus increasing the construction cost. More space is required to accommodate separate storage facilities, dedicated delivery routes, including a chemical transport elevator. Where piping containing HPM is located above a ceiling, the concealed space must be ventilated with at least 6 air changes per hour and the space cannot be used to convey air from any other area. Below grade piping must be double-contained with an outer ferrous pipe or tube.

NFPA Standards

The National Fire Protection Association (NFPA) has established design standards for laboratories using chemicals (NFPA 45) which are common reference standards for most building codes. It categorizes labs into three hazard classifications according to the types and quantities of flammable and combustible materials used. The hazard class is used to determine construction systems, size of containment areas, quantity of flammables, and emergency egress regulations. For example, high hazard, Class A labs must have two separate means of egress if they exceed 500 ft^2 in area. The NFPA standards are concerned about the location of fume hoods and compressed gas cylinder or cryogenic container storage within the facility. Even beyond NIH-CDC or NFPA standards, the design and arrangement of activities within the research laboratory can enhance and promote safety. Chemical fume hoods, the focus for many hazardous tasks, should be located away from main circulation aisles.

Eyewash Stations

People naturally go to the closest sink when there is something bothering their eyes. The reaction of certain common chemicals with the human eye is extremely rapid. The interior chamber of the eye is vulnerable within 6 to 8 seconds following contact with concentrated ammonia. An emergency eyewash outlet at the end of a short flexible hose located at lab sinks is a better alternative than a single fixed, dedicated eyewash fixture. Some labs have opted to provide portable squeezable plastic bottles in lieu of the connected plumbing eyewash with nontempered potable water. This is a cost-saving alternative. However, because safety is concerned and legal action could result in a finding of negligence because the bottle capacity is small, it is recommended that both portable and piped eyewash stations be provided (Figure 4.2). Contaminated eyes should have 15 minutes of

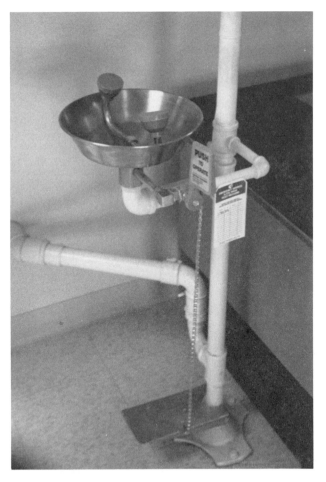

Figure 4.2.   Emergency eyewash station.

copious flushing of water at a rate of between 3 and 7 gallons per minute. This is not practical using squeeze bottles. The best units provide a multi-stream, cross flow of 65° to 75°F of potable water. For teaching laboratories, where inexperienced personnel are working, some type of eyewash should be available for every four student stations.

Emergency Showers

Emergency drench showers should be easily accessible in every laboratory dealing with chemicals, and spaced not over 75 feet on center in large open labs. Deluge showers should provide a minimum of 30 gallons of water per minute delivered at low velocity at a temperature range of 70° to 90°F. High velocity water can further damage injured tissue. Once actuated, the valve that operates the shower should require a positive action to close. In some jurisdictions, these emergency drench shower valves are required to remain in the open position for at least 2 minutes once they have been activated. These showers are not meant to extinguish clothing fires. They are for removing chemicals. (Clothing fires should be extinguished by the "stop, drop, and roll" procedure recommended by the American Red Cross.) Floor drains are not recommended under showers in laboratories. Wet vacuuming can remove the water from the infrequent use. Drains, if provided, will have their traps dry out and cause sewer gas to enter the lab. Also, vermin in the form of insects can thrive undetected below the drain cover.

Electric "Kill" Switches

Master electrical disconnect switches should be prominently located on the path of egress from the lab. To prevent inadvertent use, they should be a bright red color and mounted at eye level. These "kill" switches should disconnect all lab power in the immediate area , not exceeding 5,000 ft^2, except for lighting and life safety critical circuits such as those serving alarms and exhaust systems.

Safety Stations

A standard built-in Safety Station that incorporates emergency shower, eyewash, fire blanket, spill kit, "kill" switch, protective glasses and gloves, and portable fire extinguisher is a good design feature for the laboratory. It should be located along the path to the lab egress and identified by a large red dot on the floor surface. Another important design consideration is to locate a bulletin board near the lab entrance where safety procedures can be posted for all occupants to see.

Operating Procedures

Housekeeping and operating procedures are, of course, important in maintaining a safe research environment. Procedures and standards must be fully communicated and monitored to affect lab safety. OSHA and NIH regulations prohibit food or drinks to be within a laboratory setting. Flammable materials should be stored in fireproof enclosures. Warning signs should be posted at the lab entrances to inform the untrained visitor of the potential dangers. Each lab technician and scientist must be trained in safe practices and with emergency procedures posted.

Most open lab spaces should be under negative pressure relative to the other adjacent spaces such as offices and corridors. Lockers for storing personal items and overcoats should be kept outside the lab boundary. Adequate egress paths with emergency powered lighting should be available to areas of refuge and escape. Illumination levels must be adequate for the expected tasks. And perhaps most important, exhaust locations and fume hood face velocities need to be carefully considered and controlled.

Lab Safety Graphics

The biological safety level of the lab plus health, fire, reactivity, and specific hazards should be posted at the lab entrance. Most fire district jurisdictions require laboratory institutions to follow the National Fire Protection Association standard diamond shaped signage, NFPA 704, to warn visitors or emergency personnel (firemen) of potentially dangerous conditions. It addresses the potential for short-term (up to one hour), acute exposure to a material during handling under condition of fire, spill, or similar emergencies. There are four hazard categories to the NFPA system: health, flammability, reactivity , and special hazards. Its purpose is to provide basic information to emergency and other personnel about potential hazards within the facility. The four quadrant diamond consists of a red (top, 12 o'clock) quadrant representing the fire hazard, the yellow (right, 3 o'clock) quadrant representing the reactivity hazard, the blue (left, 9 o'clock) quadrant representing the health hazard, and the white (bottom, 6 o'clock) quadrant identifying any other specific hazard.

- Health Hazard (blue, 9 o'clock position):
    - 4 - Deadly
    - 3 - Extreme Danger
    - 2 - Hazardous
    - 1 - Slightly Hazardous
    - 0 - Normal Material

- Fire Hazard (red, 12 o'clock position):
    - Flash Points:
    - 4 - Below 73°F
    - 3 - Below 100°F
    - 2 - Below 200°F
    - 1 - Above 200°F
    - 0 - Will not burn

- Reactivity (yellow, 3 o'clock position):
    - 4 - May detonate
    - 3 - Shock and heat may detonate
    - 2 - Violent chemical
    - 1 - Unstable, if heated
    - 0 - No hazard

- Specific Hazard (white, 6 o'clock position):
  OXY - Oxidizer
  ACID - Acid
  ALK - Alkali
  COR - Corrosive
  W̶ - Use no water

 - Radiation Hazard (symbol)

 - Biohazard (symbol)

These graphics provide the visitor with a simple, easily understood general warning of the potential dangers that could take place. It indicates the degree of severity by a relative numerical rating. It is not intended to communicate hazard potential for long-term, nonemergency, occupational exposure. Obviously, considerable judgment is needed to make the appropriate numerical assignment, and to change the assignments as materials being used change. Specific reference to the NFPA 704 document should be made each time the potential risk hazards are considered for the signs.

In addition, it is good practice to have a general lab bulletin board posted near the entrance to communicate with the staff on the status of ongoing lab activities and experiments in progress. This type of communication could be crucial, should a lab emergency occur.

5

Personnel Amenities

Prevailing Culture

Attracting, recruiting, and keeping world class scientists is a major concern for research institutions such as pharmaceutical companies, universities, teaching hospitals, and industry. An important part of this strategy is maintaining a pleasant working environment. Morale is an important ingredient in work productivity, and in this respect scientists are no different than the general population, except that researchers tend to spend a larger percentage of their time at work. These people are driven in their search.

In a university or teaching hospital setting, a large percentage of the research staff is not well paid. Many are continuing their education as post-doctoral residents working under a senior mentor or lab technicians performing repetitive or less glamorous everyday tasks. Given a choice, these people would prefer to have a positive environment in which to spend their time. As an example, a high percentage of post-docs ride their bicycle to the lab because they cannot afford an automobile. These scientists understand that the few years invested in research at a relatively low income will, with effort, eventually develop into a lucrative and meaningful career.

In our society, which has a finite number of very creative and knowledgeable thinkers, the research institutions will naturally be in competition for that limited resource. Being able to provide a good working environment gives an edge to a research center and helps to ensure its long-term standing in the field.

Childcare and Fitness Facilities

Many of the younger scientists are just starting a family; more, than in years past, are women. Their needs should be addressed. Convenient child day care, exercise rooms, shower and locker facilities, safe and lockable bicycle storage space, and access to public transportation are being considered as an integral part of the laboratory program of requirements.

Group Tea Room

By far the most important amenity to provide the laboratory staff is the group tea room (Figure 5.1). Food, drink, and smoking is prohibited in the laboratory. Consequently, a break area located just outside of the lab meets a real need. These tea rooms originated at Oxford labs in England. Their capacity is usually about 20 people maximum so that informal conversations and small group discussions are facilitated. If you divide the laboratory population by 20, that will establish the optimum number of group tea rooms needed in the research building. These spaces also function as conference rooms. The typical "oasis" will have a kitchenette unit with a sink, dishwasher, microwave oven, garbage disposal, and refrigerator.

The ambiance and decor of the group tea room is better when it contrasts to the laboratory: soft materials, carpeted floor, an abundance of shelves for current journals, some upholstered chairs, a video monitor, markerboards, and projection screen. A soft couch where scientists can take a short nap is also a consideration where research activities take place around the clock. A capability for dimming the lighting is also desirable. Staff discussions are encouraged and enhanced when there is such an informal meeting space readily available.

Figure 5.1. Typical group tea room: An informal meeting area with tables, chairs, and shelves for current journals, a small kitchenette with coffee maker, microwave oven, sink, undercounter refrigerator, and garbage disposal.

Copy Center

Preparing grant proposals, writing reports and journal articles, and normal experiment documentation generates paperwork. A logical shared facility is the copy machine. Located adjacent to an informal "oasis" along the main circulation spine will encourage staff interaction and idea exchange.

Conference Center

A shared meeting space to accommodate several hundred people with good audio/visual characteristics is of particular interest in research institutions where idea stimulation is a prime goal of the workplace.

6

Bench Design

Dimensional Considerations

The lab bench is the work station for the scientific investigator. Generally, the work surface, or counter, is 36" high, as in a domestic kitchen. The worker stands or sits at a high stool. It is ideal to provide some knee space under the counter, but typically, base cabinets, small refrigerators, freezers, or storage boxes fill most cavities below the countertop. Satisfactory lighting conditions at the work surface is a paramount concern. Illumination intensity should be 90 - 100 ft-candles without glare, and it is generally ideal to have a major contribution of daylight. The position of the light source is important. The person standing at the bench should not work in his/her shadow. Like the artist studio, space with high ceilings and a generous window area is highly desirable for bench work (Figure 6.1).

The minimum bench width for a single researcher is 5 feet. However, with the growing use of computers and other bench top equipment within the lab, a more realistic width is 7 lineal feet per person. Peninsula-type bench layouts with access from one end only should not exceed 15–18 feet for safety reasons, or space for a maximum of three people. The demand for storage space within easy reach of the bench worker necessitates shelving above and in cabinets below the bench top (Figure 6.2). Open shelving above the bench counter is more popular than shelf cabinets with sliding or swinging doors, unless used for storage of sterile glassware, instruments, or drugs where locks may be required. Glass doors are often preferred over opaque door material for wall

Figure 6.1. Lab bench peninsula with researcher desk at the end adjacent to the window.

Figure 6.2. Lab benches before and after occupancy. Generally, laboratories are cluttered environments. Note the rack above the sink for hanging glass and plastic columns. There are never enough shelves!

Table 6.1. **Practical Bench Dimensions**

Type	Bench Height	Seat Height	Min. Depth Kneehole	Min. Vert. Floor to Under Counter
Sitting Only	27.5"	16.75"	22.5"	26.125"
Women (average):	33.375"	24.625"	22.5"	31.5"
Men (average):	35.5"	26.5"	22.5"	33.375"

mounted cabinets. The minimum practical dimensions for bench sinks are 18" × 18" × 12" deep (Table 6.1).

Modularity

Modular base cabinets should be considered for adaptability to future changes. A nominal two-foot base cabinet width (23 5/8") works well for flexibility with movable units. Bench base cabinets are available in hardwood with a tough urethane finish, steel with baked enamel finish, and wood particle board with laminated plastic finish. Researchers usually prefer an equal distribution of under counter drawer and shelf units. If possible, the bottom drawer should be a standard file drawer for 8 1/2" × 11" papers. Because of the premium for counter space, most researchers like to have a pull-out writing board just under the counter at the top of the 2' wide drawer cabinet.

Flexibility

Another popular approach to bench design is to separate the cabinetry from a more permanent fixed shelf structure and enclosure for the services such as DI water, gas, lab water, cup sinks, vacuum, power and data connections. This allows greater flexibility in placing the bench cabinets at any point along the service enclosure. A sink with its drain and faucet connections is the only relatively fixed cabinet element with this design concept. It also allows changes to be made in the bench top height, but it does mean there will be a joint between the service enclosure and the bench top which is supported from the movable cabinets. Most scientists believe that flexibility is far more important within the lab support zone as compared to the more generic bench area.

Some lab bench designs carry flexibility to a further extreme by providing a cantilevered structural framework from the service enclosure to support the bench top. With this arrangement it is very easy to relocate, or even eliminate base cabinets altogether.

Drains

The one crucial consideration for wet bench design is the location for the drain and vent stack serving a sink. Since fluid flow is gravity dependent, the lab waste pipe must pitch downward, a minimum 1/4" per foot by code. There is a limit to the lateral distance the drain pipe can run before it must penetrate the floor. The position of these penetrations and associated vent stacks is a primary consideration for integration with the other design components such as the building structure, walls, and placement of functions on adjacent floors. Except for the lab drain, all other bench services can come horizontally, from overhead, or both. The gravity dependent drain piping location constrains lab bench design flexibility and must be carefully considered in positioning sinks and benches in a laboratory.

Work Surface

Choosing an appropriate countertop material for the lab function is critical. Countertops are available in a variety of materials with a wide range of initial costs. Epoxy and epoxy impregnated sandstone are perhaps the most versatile surfaces for use with acids and solvents. Of these, epoxy has the shorter lead time for quick order situations. Chemical-resistant laminated plastic provides an acceptable work surface for many laboratory tasks at much less initial cost. Stainless steel counters are more appropriate where surgical procedures are performed on animals or where there may be a danger of radioactive spills. In electronic labs, wood butcher block bench tops should be considered where soldering and cutting work is performed. The work surface should not have a gloss that would cause a reflected glare to the worker. A light colored bench top is superior to a darker color to reduce contrasts and reduce eye fatigue.

Utilities

Services must be provided as appropriate for the tasks to be performed at the bench. For wet bench work, options include a plug mold for 110 vac power mounted at the bottom of the first over-counter shelf, vacuum and gas turrets at about 5-foot horizontal spacing. The plug mold should have one outlet per lineal foot and one duplex under-counter outlet at one knee space for a potential under-counter refrigerator or freezer. Two separate duplex outlets are appropriate to provide at the end of "dry" peninsula benches. Lab sinks should have nonpotable hot and cold water and de-ionized (pure) water, plus tempered potable water serving a flexible hose emergency eyewash spray. Cup sinks are often provided in wet bench tops for disposing of small quantities of liquids. Space needs to be provided in the concealed space behind the base cabinets for the plumbing waste piping that is gravity dependent. Communication outlets are becoming more necessary with so much new countertop computer-based scientific equipment (Figure 6.3).

Back-to-back peninsula benches often have a sink located at the freestanding end with a column rack and glassware pegs mounted above the sink. This sink should be made from stainless steel or epoxy coated resin at least 16" deep by 18" to 24" in horizontal dimensions. A small vertical chase for services behind the peninsula sink can contain the plumbing vent stack, conduit for electric power, and piping for gas, vacuum, and pure water systems. Peninsula benches are typically 60" wide and between 10' and 11'6" on centers.

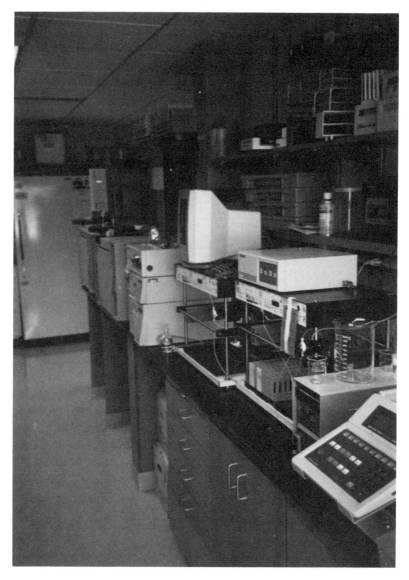

Figure 6.3. Benchtop equipment is ubiquitous. Electronic devices for measuring, sensing, and recording data are replacing the manual lab tools formerly found at the work surface.

It is a good idea to provide at least one dedicated, conditioned power outlet at each bench for critical power needs. Sit-down, 29" high desk space should be close to the benches for writing notes and entering computer data. This desk space can be built-in 4' to 5' wide with laminated plastic counter, knee space, nearby tackboard back panel, convenient two-drawer file cabinet, shelf space, and cove lighting above the work surface.

Shelving

There are never enough shelves in a laboratory to provide storage space for the items needed. The high shelves above the bench top should have a small (1/4") safety lip at the front edge to inhibit accidental spills from containers sliding forward when the shelf contents are being moved. Modular heavy duty 36" long × 12" deep adjustable shelving with turned-up ends should be provided wherever possible. Some shelf modules may have to be 30" long to use all of the available wall space efficiently. Benches for electrophoresis work need deeper shelving (16") with a 1" gap at the rear for the power chord pigtails from the equipment that will be placed on these shelves.

Caveats

A few words of caution: researchers have changed their habits over the years, but they seldom realize it. For example, the use of hot plates has virtually eliminated the need for Bunsen burners with their dependence on having natural gas provided at the bench. Providing gas piping to all benches requires a substantial investment. Limiting the scope of natural gas service to a few benches only supplied from a replaceable tank may be more appropriate. Cup sinks are another feature that involves a lot of cost with questionable benefit when vacuum outlets are also available for disposing of small quantities of liquid. Before deciding on the right countertop material, consider how much of the bench top is devoted to providing a platform for equipment. Light gray laminated plastic costs considerably less than black epoxy impregnated sandstone and it provides less contrast and better viewing conditions for the scientist. Finally, remember that flexibility usually comes with a cost premium. Lab benches seldom change as much as lab support space, so try to prioritize the initial cost premiums to best serve the future lab needs.

7

Ergonomics

Sight

Research is a learning activity, receiving and processing information. Scientists tell us that 90% of the information we gather comes through vision. The eye is naturally attracted to light sources within the parabolic field of vision. The eye is a natural light-grading detector that tends to fixate on contrasts. It follows, then, that designs that contribute to a learning environment, to assist people in performing their work efficiently and effectively, will provide adequate levels of illumination, reduce opportunities for contrasting surfaces and glare, and keep the potential areas of focus centered within the natural vision field while providing a high degree of physical comfort.

Focus

Learning environments must be free of distraction in order to enhance the ability of the scientist to remain focused and concentrated. Background noise, echo, uncomfortable temperature, stale air, glare, noticeable air flow, unusual odor, stressed muscles, and light contrasts within the field of vision are negative influences to focusing on task. The laboratory designer must purposely manipulate spaces, surfaces, light sources, and environmental system components in such a way as to control the "noise" to signal ratio. Undesirable environmental influences (noise) must be minimized, while focusing on work activity (signal) is promoted. This should be a major objective of a designer of spaces for learning.

The video display terminal (VDT) is becoming a critical area of focus in the age of nanoscience where computer displays are replacing images of microscopy. The finest computer image in the world will be of little benefit if it is not coordinated with room design and seat selection. A viewer must be relaxed and comfortable over long periods of screen focusing. Glare reflections must be nonexistent, especially from the screen surface. The background illumination level should not be too high as to "wash out" the displayed image, and the luminance level on the screen must be bright enough for a desirable range of brightness contrast for the projected image. Avoid distracting background colors such as saturated reds or blues which can increase eye fatigue. Keep both keyboard and VDT low to avoid neck muscle strain. Frequent chin-lifting head movements and extended chin-up positions lead to discomfort and muscle fatigue. Select an adjustable fully ergonomic chair that can accommodate the various user body sizes with a concave seat pan cushion for proper weight distribution and low back support. Avoid small video screens that do not dominate the cone of vision; symbol size is an important factor in understanding the image.

Sound

Background noise levels, especially at contrasting frequencies, contribute to distracting environmental conditions. An extremely quiet space is nearly impossible to achieve and would probably seem so unusual as to be uncomfortable. As a practical matter, some background sound is inevitable. Air passing through a diffuser, fans, footsteps, traffic noise from outside, computer keyboards, etc. all contribute to the milieu. There are several methods of dealing with the sound conditions of a workspace.

First, isolate or mute the disturbing noise sources. This is done through containment. Sound is a wave action at different frequencies. Containment is achieved by the mass and limpness of the container wall and by sealing any penetrations and joints that would allow the passage of air. Second, absorb the sound with inefficient sound-reflecting surfaces. This is more difficult in a wet laboratory where the need for hard, washable surfaces has a high priority. Third, locate objectionable sound sources remote from work spaces. Sound waves diminish with distance traveled. And finally, where good listening conditions are not critical, introduce "acoustic perfume," background music, or "white" sound at the full spectrum of the hearing frequencies that will mask the disturbing noises.

8

Chemical Fume Hoods

Definitions

The purpose of chemical fume hoods is to provide personnel safety. It is a ventilated enclosure where hazardous materials can be handled. The chemical fume hood prevents contaminants from escaping into the laboratory. This is accomplished by drawing the fumes within the work chamber away from the worker so that inhalation and contact are minimized. The concentration of contaminants in the worker's breathing zone must be kept as low as possible, and should never exceed the threshold limit value (TLV) for the materials being handled.

The containment of contaminants is based on the principle that a flow of air entering at the hood face, passing through the enclosure, and exiting the exhaust port will prevent the escape of airborne contaminants from the hood into the room. The degree to which this is accomplished depends on the design of the hood, its installation, and its operation. There are three fume hood application levels:

- *Class A*: Materials of extreme toxicity or hazard, such as volatile carcinogens, perchloric acid, tetraethyl lead, beryllium compounds, metal carbonyls, etc. (for materials with TLV less than 10 parts per million). Should also be used where heat above 150°F is introduced with the potential for igniting volatile materials in the hood chamber. Recommended average face velocity of 125–130 FPM, with minimum at any one point of 100 FPM.

- *Class B*: Normal laboratory usage with materials of average toxicity or hazard (for materials with TLV between 10–500 parts per million). Recommended average face velocity of 100 FPM, with minimum at any one point of 80 FPM.
- *Class C*: Materials of low toxicity or hazard or for operations creating nuisance dusts and fumes (for materials with TLV greater than 500 parts per million). Recommended average face velocity of 75–80 FPM, with minimum at any one point of 50 FPM.

In addition, there are several basic hood designs:

- *Conventional*: An enclosure with an interior baffle that controls the pattern of air movement through the hood and a transparent movable panel, or sash, set across the hood entrance. Hood performance is dependent on the opening sash position. Closing the sash disrupts the air flow which can cause damage to fragile experimental apparatus (Figure 8.1).
- *Bypass*: Similar to the conventional hood, except as the sash closes, air bypasses the sash closure and the face velocity is kept constant. This is a better hood for fragile types of work. The bypass type of hood should be used for Class A and Class B applications (Figure 8.2).
- *Make-Up Air*: Fume hoods that can provide an auxiliary source of filtered and heated, but not cooled, air immediately above the hood face. This can reduce the loss of cooled building air by as much as 75%. These auxiliary air hoods do require a separate supply fan and duct system to each such hood, which increases initial cost and takes up valuable building space. This type of hood is available as a conventional hood or a bypass hood. The make-up air hood may be the most economical for air-conditioned labs. In evaluating the cost benefits of using the make-up air hoods, the initial cost reduction for smaller capacity room air cooling system and annual operating expense savings must be measured against the cost of additional ductwork and fans (Figure 8.3).

Many special purpose hoods are also available. Among the most common special purpose hoods are:

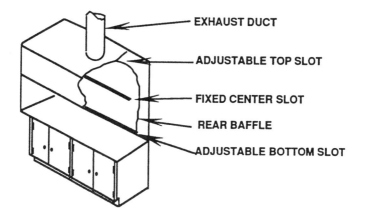

EXHAUST DUCT

ADJUSTABLE TOP SLOT

FIXED CENTER SLOT

REAR BAFFLE

ADJUSTABLE BOTTOM SLOT

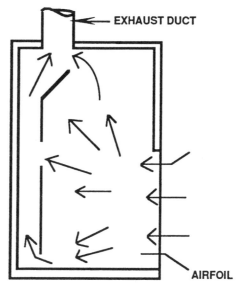

EXHAUST DUCT

AIRFOIL

Figure 8.1. Conventional chemical fume hood.

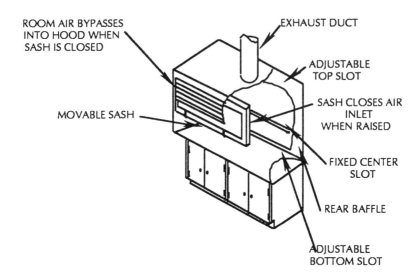

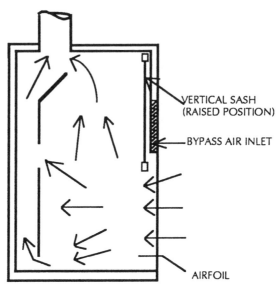

Figure 8.2.   Bypass chemical fume hood with vertical sash opening and bypass air inlet.

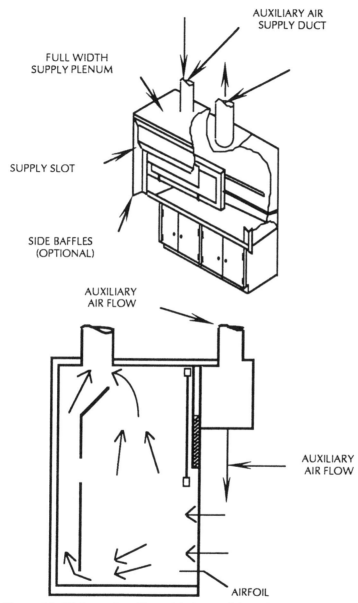

Figure 8.3. Make-up (auxiliary air) chemical fume hood.

- *Radioisotope*: Class A, bypass type of hood with integral seamless bottoms and coved interiors to facilitate decontamination. These hoods are reinforced to support lead-shielding bricks. To conform to radiation safety guidelines, a particulate air sampler pump must continuously monitor the exhaust effluent for hoods where iodination work is performed or irradiated isotopes are handled.
- *Perchloric Acid*: Class A, bypass type of hood with integral bottoms, coved interiors, and a drain. Wash-down features are incorporated since the hood and duct system must be thoroughly rinsed after each use to prevent the accumulation of reactive residue. An air flow monitor is also recommended for the exhaust effluent. The hood interior is constructed of relatively inert materials such as type 316 stainless steel, ceramic-coated material or PVC. These hoods should be dedicated for perchloric acid work only and a warning label attached.

Research technicians often use the term "hood" for other laboratory equipment that is similar in appearance to a fume hood, but are not considered fume hoods. These nonhoods include:

- *Biological Safety Cabinet*: (BSC) Special safety enclosure used to handle pathogenic microorganisms. It contains a HEPA filter and fan to recirculate room air. Much tissue culture work can be performed safely in biological safety cabinets that do not require a direct exhaust to the outside.
- *Laminar Flow Cabinet*: Clean bench or biological safety cabinet that incorporates a smooth directional flow of air to capture and carry away airborne particles.
- *California Type Hood*: A rectangular enclosure used to house distillation apparatus that can provide visibility from all sides with horizontal sliding access doors along the length of the assembly. The enclosure, when connected to an exhaust system, will contain and carry away fumes generated within the enclosure when the doors are closed or when the access opening is limited.
- *Canopy Hood*: Suspended ventilating device designed to exhaust only heat, water vapor, and odors.

- *Glove Box*: Total enclosure used to confine and contain air sensitive or water reactive hazardous materials with operator access by means of gloved portals or other limited openings. These are not permitted in BL-1 or BL-2 laboratories.
- *Table Top Hood*: A small, spot ventilation hood for mounting on a table that is normally vented down through the table top. Used primarily in educational laboratories to control noxious fumes.

Placement Considerations

The location of a chemical fume hood within a lab is important. Generally, it should not be adjacent to the path of egress or heavy traffic area since explosions or fires are more likely at the hood, due to the activities being performed. It should be remote from cross drafts and air currents that could influence the hood face velocity (Figure 8.4). Obviously, the fume hood should relate closely to the vertical shaft that contains the exhaust ducts to minimize horizontal duct runs.

The person standing at the hood should avoid quick movements with the hands or body that cause disturbing air currents. Adequate space (minimum 48") should be clear in front of a hood as a dedicated zone for the hood operator.

Energy Consumption

The spacing frequency and types of chemical fume hoods (both present and future) to be housed within a laboratory determine, to a large extent, the basic parameters for the HVAC systems. The amount of make-up air, exhaust capacity, size of the ductwork, power needs, boiler capacity, and refrigeration needs are primarily determined by the number and type of fume hoods to be provided. The size and, to some extent, location of utility service shafts are largely a function of the fume hood characteristics. They also have a most significant influence on amount of energy consumed and subsequent level of operating cost. Design engineers

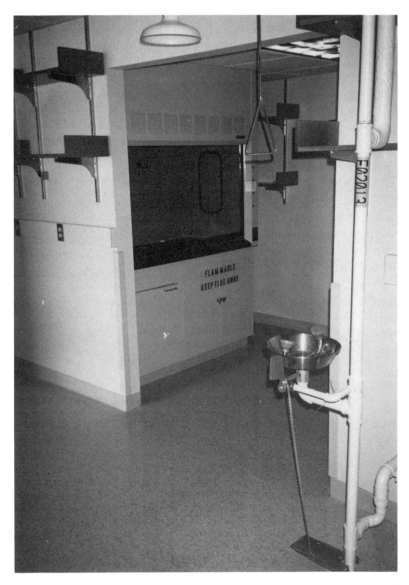

Figure 8.4. Chemical fume hood alcove with emergency eyewash and shower in the foreground.

must determine the quantities of air to be exhausted to predict the amount of make-up air that must be conditioned to provide acceptable environmental conditions.

When considering life-cycle costs, fume hoods represent an enormous investment. They waste (exhaust) filtered, air-conditioned building air continuously to the exterior 24 hours every day of the year. Sizes of supply air fans, ductwork, chillers, exhaust fans, power supplies and emergency generators are greatly influenced by the number and size of laboratory fume hoods. Generally, fume hoods create exhaust needs well beyond the requirements for normal room ventilation. Fume hoods are responsible for a large portion of the energy consumed at a laboratory facility.

Every 300 cubic feet per minute (CFM) of air exhausted requires approximately one ton of refrigeration. To determine the CFM required for a fume hood, multiply the sash opening area in square feet times the face velocity (FPM).

For installations incorporating an individual exhaust fan for each hood, a two-speed fan can be interconnected to the face sash at the hood so that when the sash is down the fan can be on low speed, thus saving energy without compromising safety.

Exhaust Systems

Fume hood exhaust ducts should be as short, direct, and vertical as possible to the terminal discharge. The exhaust fan should pull the air from the terminal end rather than push the air from the fume hood. (If the exhaust air is pushed through the duct it could escape at any leaking joint within the building.) The duct interiors should be smooth, impervious, and cleanable. Round duct sections are preferable over rectangular sections because they have less friction loss or resistance to flow. Any horizontal duct runs should pitch back toward the fume hood since condensing vapors could be corrosive. Dampers should not be installed in fume hood exhausts (most building codes exempt fume hood exhaust ducts from requirement for fire dampers at rated wall assemblies).

Stainless steel fume hood exhaust ducts are required in some jurisdictions. However, if permitted, 26-gauge spiral galvanized steel or PVC duct material works satisfactorily if the exhaust duct is exposed or easily accessible. These are less expensive materials. If the fume hood exhaust ducts are to be permanently enclosed and inaccessible, or if highly corrosive materials will be used in the hoods, stainless steel ducts should be used. Besides stainless steel, other materials to be considered for acidic, toxic, or pyrophoric fumes include epoxy-coated steel, polypropylene, Teflon, or other corrosion-resistant materials.

Scrubbers are sometimes necessary to remove aggressive substances from the waste gas stream, such as perchloric, sulfuric, or hydrochloric acids. Fume hood scrubbers create a high velocity spray jet mist, or finely atomized liquids, which results in a separation performance between the gas and liquid substances. The scrubbing liquid is selected to be the most appropriate for the type of hazardous material to be separated. These scubbers have become extremely sophisticated in that they can include sensors connected to a computer which can automatically analyze and alter the liquid spray material to correspond to the hazardous substances present in the exhaust air stream.

It is safer to have a separate duct, fan, and scrubber for each hood, although there are many installations where several hoods have been ganged, or manifolded, together to a common exhaust fan. In some instances this makes good sense, provided there is no danger of mixing effluents that could cause an explosive hazard, or make it impossible to identify a particular hood that is releasing excess radionuclides. In any event, the HVAC engineer expert in the design of these systems needs to be consulted for all fume hood exhaust conditions.

For radioisotope hoods where ducts may be dismantled for decontamination, flanged neoprene gasketed joints with quick disconnect fasteners will provide minimum time exposure to the decontamination personnel. A sampler pump is required within the exhaust duct or fan where radioactive I-125 isotopes are present to monitor the effluent. For those installations where high volume iodine work is anticipated, a charcoal filter should be built into the exhaust duct, where they can be easily maintained.

Perchloric acid fume hoods should have nonmetallic stainless steel exhaust ducts. Perchloric acid deposits in ductwork can become a major explosion hazard. Internal water spray systems for periodic washing of the duct interior surfaces are mandatory for perchloric acid exhausts. Joints in these exhaust ducts should be welded and ground smooth.

Hood Sizes

The most common outside lengths for fume hoods are: 4', 5', 6', and 8' with a working surface depth about 25" to 26". The most popular bench hoods have a work area height of 34" to 48" available. The volume of the work space within the hood should be determined by needs. If a hood larger than needed is used, initial, energy, and operating costs will be wasted. Hoods should be evaluated by actual space needs for apparatus and materials. The hood should not be used as a storage enclosure for nonessential corrosive, toxic, or flammable materials. Not only does such storage jeopardize hood performance and create hazards, it also takes up valuable space.

Hood Materials

Other than cost, there are three basic considerations in determining the proper materials for the fume hood work area:

- Nature of hood effluents
- Ambient temperature
- Flame and smokespread rating.

Fume hoods, exhaust ducts, and fans are subject to attack from effluents by corrosion, dissolution, and melting. Effluents vary in temperature and are classified as organic or inorganic chemical gases, vapors, fumes, or smoke. They can be acids, alkalis, solvents, and oils. An evaluation of expected effluents and potential decontamination materials is a necessary step in selecting the most appropriate hood lining materals.

Although expensive, type 316 stainless steel should be selected for radioisotope and perchloric acid applications. Cement-fiber board and fiberglass are the least expensive hood liner materials. The cement-fiber board has better heat and flame resistance, but the fiberglass has better stain and moisture resistance. Other available hood lining materials are epoxy-coated steel, epoxy resin, polypropylene, and polyvinylchloride (PVC).

Optional Features

Air flow monitors for the hood face are available with high and low flow warning signals for laboratories that handle particularly critical materials that require constant face velocities.

In order to provide a proper level of illumination within the fume hood work area, light fixtures are recommended. They are available for external or internal mounting. A transparent, impact-resistant shield should be provided for externally mounted fixtures. Interior mounted lights should be either vapor proof or explosion proof. In all cases the light switch should be outside of the hood. Likewise, any electrical receptacles should be mounted on the hood exterior.

Service fittings, if required, such as gas, water, vacuum, air, etc., should be installed to allow the connection of service supply lines through an access panel in the hood. The control fixtures for these services should be external to the hood, clearly identified, and within easy reach. If water is supplied, a cup sink drain should be included at the rear of a dished corrosion-resistant work surface.

Depending on the hazard level of the effluent, and the degree of pollution abatement desired, exhaust filtration may be necessary. For convenient handling, disposal, and replacement with minimum hazard to the personnel, exhaust filters should be located:

- outside of the laboratory area, unless it is an integral part of the hood
- ahead of the exhaust fan
- in space that provides free, unobstructed access
- positioned at a convenient working height.

A damper is often added to HEPA filtered exhausts to balance air flow, because these filters vary and change in resistance as they fill. Wet collectors or adsorption systems (e.g., activated charcoal) can be used for removal of gas-phase toxic or odoriferous pollutants in the effluent.

9

Focused Work Spaces

Many investigative tasks and techniques cannot be performed efficiently at the benchtop. Equipment with specific environmental conditions or services requires customized work space designs. Controlled ranges of light intensity or spectrum, humidity, temperature, ventilation, pressure, vibration, sound, and cleanliness levels all contribute to the space characteristics and design details (Table 9.1).

Tissue Culture

Culturing tissue is a major component of biomedical investigation—the growing of bacteria or other microorganisms in a specially prepared nourishing substance, or media. This work is conducted in a dedicated separate space from the generic lab benches (Figures 9.1 and 9.2). The room should be under positive pressure with respect to the surrounding spaces. Obviously, this means that the surrounding partitions must be full height to the deck with all penetrations tightly sealed. A nine-foot dimension provides the most efficient tissue culture room width for the required biological safety cabinet, double upright carbon dioxide (CO_2) incubators, microscope table, refrigerator, sink, shelves, and cabinets. The flooring should be seamless with a flashed integral base for easy cleaning. A standard lay-in acoustic panel ceiling is generally acceptable; however, some researchers prefer the smooth clean room type panels. Illumination levels should be at least 80 foot-candles in the room. Separate 115 V, 20 amp circuits should be provided to each biological safety cabinet for the fan and light.

Table 9.1.   Laboratory Space Characteristics

Type Space	Floor	Base	Wall	Ceiling	Door	Lighting	Pressure	Remarks
Open Bench	VCT	R	P FH	AC	SC OP	FL	N	
Tissue Culture	SV	S I	P FH	AC	VP SC	FL	N	Panic Bar In-Swinging
Clean Room	SE	S I	E FH	HEPA	AL SC VP	FL	+	Gowning - Laminar Air Flow
Electrophoresis	VCT	R	P	AC	VP	FL	0	17" deep shelves with 1" gap at rear
Microtome	VCT	R	P	AC	VP	FL	0	Min. Air Flow
BL-3 Work	SV	S I	E FH	E	AL VP SC	FL	N	Autoclave - Gowning - Hand Washing "Hot" Lab
Isotope Work	SV	S I	SP FH	AC	OP SC	FL	N	
Electron Micro.	VCT	R	P FH	AC	OP SC	IN D	0	Adjacent Darkroom
Fluor. Micro.	VCT	R	P	AC	IU OP	IN D	0	30" H. Bench
Cold Room	SV	S I	M FH	M	VP SC	FL	0	Vapor Barrier
Darkroom	VCT	R	P FH	AC	OP IU	IN D R	0	Dark Walls Light Tight

	Floor	Base	Walls	Ceiling	Door	Lighting	Room Pressure	
X-O Mat	VCT	R	P FH	AC	OP IU	IN D R	0	Indirect Drain
Glasswash.	SE	S I	E FH	E or PAC	VP DA SC	FL	N	Waterproof Fixtures
Animal Cage	SE	S I	E FH	E	SC VP	FL	+	Waterproof Fixtures
Infect. Animal	SE	S I	E FH	E	SC VP	FL	N	Waterproof Fixtures
Laser	VCT	R	P FH	AC	OP IU	IN D	0	Vibration Sensitive
Electrophy.	VCT	R	P	AC	IU OP	IN D	0	Dedicated Ground
Procedure	SE	S I	E FH	E	OP SC	FL IN	+	Stainless Steel Work Surfaces

LEGEND:

FLOOR
SV = Seamless Vinyl
SE = Seamless Epoxy
VCT = Vinyl Composition Tile

BASE
S I = Integral with Seamless Floor, Coved
R = Resilient, Coved

WALLS
E = Epoxy Coating, Vapor Barrier
P = Washable Paint
SP = Plexiglas Shielding over Washable Paint
M = Painted Metal
FH = Full Height to soffit of structure above

DOOR
AL = Air Lock
VP = Vision Panel
SC = Self-Closing
I U = In-Use Light
DA = Double Acting
OP = Opaque

LIGHTING
FL = Fluorescent
I N = Incandescent
D = Dimming
R = Red Light

ROOM PRESSURE
+ = Positive
N = Negative
0 = Neutral

CEILING
AC = Acoustic Panel
PAC = Plastic Coat. Acoustic Panel
E = Epoxy-Coated Gypsum Board
HEPA = Clean Room Air Filters
M = Painted Metal

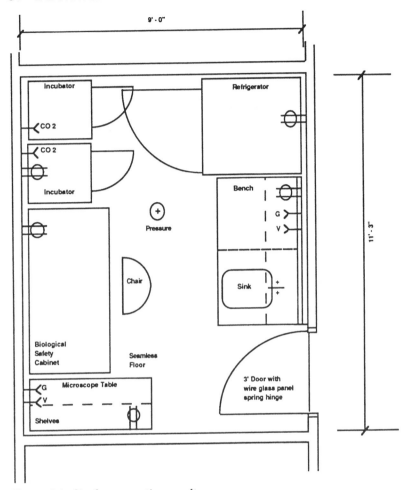

Figure 9.1. Single-person tissue culture room.

Mold is the enemy of tissue culturing; therefore, the refrigerator drain pan and the incubators must be kept clean. The concealed space above a suspended ceiling must also be free from mold growth. If possible, drain pans for suspended fan coil units should be isolated or contain some Lysol or other disinfectant to prevent mold growth. Because of the ever-present possibility for mold contamination, most scientist prefer to have a suite of several tissue culture rooms rather than a single multi-station facility. Because the scientists can

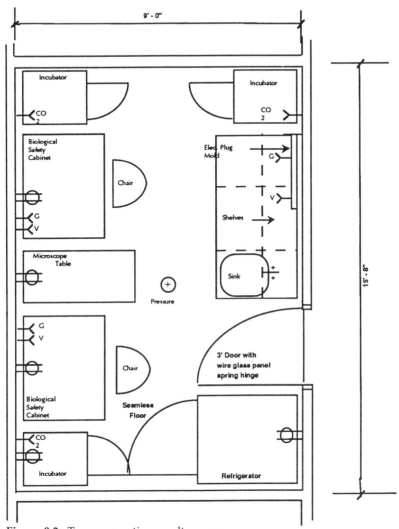

Figure 9.2. Two-person tissue culture room.

spend considerable time in this relatively small room, they prefer to have a window, which is usually in the door, since the wall area is full of equipment and shelves. The minimum three foot by seven foot in-swinging door must be self-closing, but spring hinges are acceptable and less costly compared to a closer. A panic bar device mounted across the out-

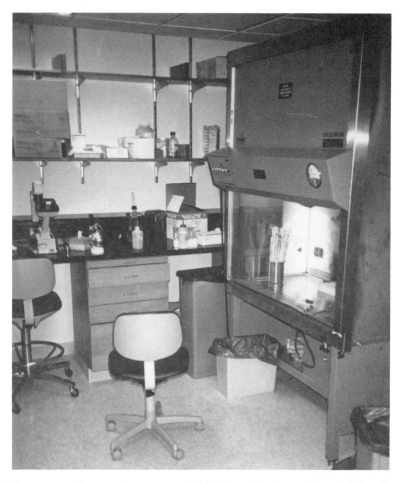

Figure 9.3. Tissue culture room with biological safety cabinet, lab bench, and microscope.

side face of the door allows the researcher to open the door with his/her body when both hands are carrying media material from the lab.

The tissue culture media preparation work is often conducted at a counter in the glass washing room, or sometimes in a dedicated central media prep space that serves as an ante room to several tissue culture rooms. It is best not to place the CO_2 tanks that serve the tissue culture incubators within the room. These are better positioned near the delivery access and manifolded together with active and standby groups with automatic changeover when the active tanks become empty. This arrangement permits a continuous gas supply to the incubators and recharged tank replacements without disruption.

Microscopy

There are many types of microscopes or optical instruments which can magnify very small objects so that their structure or texture may be examined. Simple, compound, and binocular microscopes are generally used under normal laboratory conditions, mounted at the bench, table, or low counter. No special environmental conditions or services are required for satisfactory operation.

The fluorescent microscope is generally located in a dedicated internal room. The lighting intensity needs to be adjustable, with incandescent general lighting fixtures with dimming controls within the operator's reach. The counter height for mounting the fluorescent microscope is 29", or normal desk height. A duplex 110 vac outlet above the counter is also necessary. The counter for one person needs to be only 42" wide by 30" deep and can be laminated plastic. The fluorescent microscope room should have a door without a vision panel so that total darkness can be achieved when required. Generally, the fluorescent microscope room is quite small (25 ft^2) and serves only one person for a short period of time. It is often a shared facility among different labs. However, one such room per floor should provide a 60-inch turning radius space within the room for personnel in a wheelchair.

The confocal microscope can magnify living tissue. It is basically a powerful computer linked to a 29" high workstation and requires room air-conditioning to offset the heat output from the equipment. The same darkening conditions are required as with the fluorescent microscope room, but the room is larger (about 100 ft^2) to accommodate the equipment space. It is good practice to provide an "in use" light outside this room so that there will not be accidental disruptions of the operator's work. Because of the value of the equipment, the confocal microscope room should be lockable.

Scanning electron microscope (SEM) and tunneling electron microscope (TEM) also require dimming the general room incandescent illumination. The single large SEM or TEM is centered in a larger room (180–225 ft^2) with supplementary chillers and electrical equipment housed in an adjacent closet. The SEM or TEM room usually has an adjacent darkroom for developing films of the magnified images. These highly sophisticated and expensive microscopes include a tall column which usually requires at least a 9' 6" high ceiling or coffered recess in the ceiling above the instrument column. These microscopes have an attached console. The operators may spend long periods of time in the microscope room; therefore, a telephone should be conveniently available. Independent access to the supplementary chillers and electrical equipment is a good idea so that maintenance can take place remote from the expensive microscope. Besides power, chilled water supply and return piping connections are required at the chiller unit. Sometimes nitrogen and compressed air connections are required at the microscope units. These sensitive instruments must be isolated from potential sources of vibration such as elevators, loading docks, and service corridors.

Electrophoresis

Electrophoresis is the movement, under the influence of an electrical field, of electrically charged particles suspended in a fluid; or in medicine, the introduction of a liquid into body tissue by means of an electric current. These tasks generally involve very high voltages with potential hazards to the scientists. Although this work is often performed at the open lab bench it is safer if isolated from other laboratory procedures. A dedicated electrophoresis work room consists of a

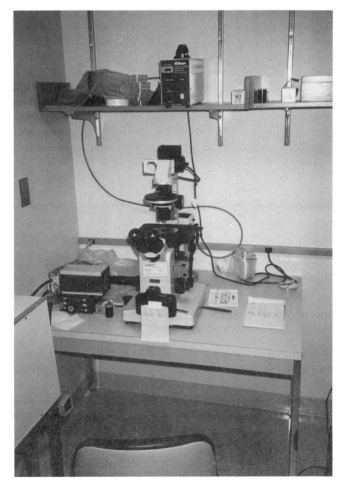

Figure 9.4.   Fluorescent microscope. Locate in room with incandescent
lights with dimming capability and light-tight door.

30″ deep counter or bench with storage cabinets below and
deep shelves (17″) above and a power plug mold mounted
on the wall above the counter. The deep shelving is for
mounting of electrical equipment such as transformers,
pumps, and gel dryers. A continuous 1″ gap should be provided at the back of the shelves for the wiring pigtails so
they do not dangle in front of the shelves. If a door is pro-

vided to the electrophoresis space, it should have a large vision panel so that the worker in the space can be observed in the event of an electrical accident. A red "panic" switch should be nearby to disconnect the power to the electrophoresis area in the event of an emergency condition. With the advent of new low voltage, low amperage benchtop electrophoresis equipment, the need for a dedicated room for safety is becoming less important.

Electrophysiology

This is the study of electrical phenomena produced by or in living organisms. Rooms for this work should be capable of darkened conditions; windows will need blackout drapes. A good dedicated electrical ground is also required for the sensitive electronics gear.

Microtome and Micro-Injection

Microtome: an instrument for cutting thin sections of tissue for study under the microscope. Micro-Injection: placement of a foreign gene into a single or multi-cell live embryo. Both of these procedures are especially delicate. They cannot tolerate distractions such as noticeable air flow from diffusers or from moving objects, such as people walking by. Good lighting and quiet air are essential considerations in designing spaces for these activities. Work is performed at low (29") benches.

Glasswashing and Media Preparation

These two very different functions can share the same space. Autoclaves, glasswashing machines, dryers, and electric steam generators are located in these rooms. The seamless and waterproof floor within the equipment zone should have a continuous surrounding curb and a floor drain. High humidity conditions make it desirable to have self-closing double-acting doors with sweep rubber astragals. A deep

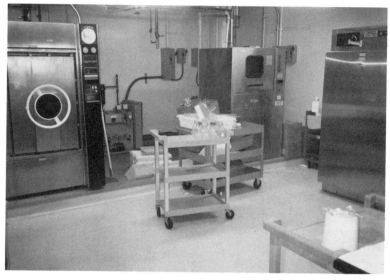

Figure 9.5. Glasswashing room. Autoclave, glasswasher, and electric steam generator mounted in waterproof zone behind 6-inch high curb. Dryer unit mounted outside waterproof zone on seamless vinyl floor.

stainless steel sink for washing columns and pipettes and a 6′ long bench for preparing the media used in the tissue culture work are often included in the glasswashing area.

The washer is usually on a separate 208 vac, 30 amp, 3 phase circuit with the oven on another 208 vac, 30 amp, single phase circuit. A steam generator for the autoclave will be required if house steam is not available. The steam generator can have a load between 45 to 80 KW, depending upon capacity. The autoclave will also need two 115 vac terminal boxes. A typical 275 ft² glasswashing room will have about six general 115 vac outlets on two circuits, with waterproof enclosures.

Darkrooms

Darkrooms require an interlocked sally port vestibule, or red warning occupied light to protect the darkness condition during critical film development work. It needs a two-compartment sink with a generous drainboard, bench and cabinets for chemicals and paper storage, wall shelving, and a small refrigerator for film storage. The darkroom light should have a filter for processing X-ray film where appropriate. A single incandescent light (not fluorescent) for general maintenance purposes should be provided in addition to the darkroom light. The switch for the normal incandes-

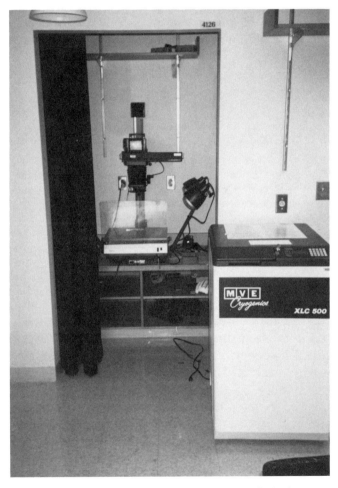

Figure 9.6.   Dark booth. A small (3' × 4') closet with darkening curtain and a Polaroid camera mounted on 24 inch high by 24 inch deep laminated plastic counter with shelves above and below.

cent fixture should be mounted above, 60 inches off the floor with a coverplate to prevent accidental operation. Two 208 vac, 20 amp, single-phase circuits are necessary for special film processing equipment, and if an ultra-cold freezer is to be located in the darkroom, a 208 vac, 20 amp, 3-phase circuit will be necessary. In addition, two 115 vac circuits serving six duplex outlets at 48 inches above the floor and one 115 vac circuit serving three duplex outlets at 16 inches above the floor are recommended to serve potential equipment often positioned in the darkroom. A silver recovery unit mounted under the sink is often required to prevent wastewater contamination from the film developing process. The Americans with Disabilities Act (ADA) places unique access challenges on the darkroom designer.

Environmental and Test Chambers

These special rooms can satisfy many different research needs. Work efficiency should be carefully considered since temperature conditions are outside the normal comfort zone for personnel. Areas of increased activity such as work surfaces and sinks are best located near the exit door. A working cold room is an extension of the laboratory and usually includes a bench with gas and vacuum outlet, sink, plug mold, and "monkey bars" for column work. They can be fabricated to provide a constant temperature within ±0.5° of 4°C which is necessary for chromatography work. Nematodes (worms) and insects are kept in unique environments for some biomedical labs. Environmental rooms can function as temperature extreme test chambers for equipment under development. Less expensive walk-in meat locker-type freezers without the precise temperature uniformity are used to store blood samples and animal parts prior to disposal. Dry rooms with relative humidity as low as 2% are important with moisture-sensitive and hygroscopic products. Some areas of pharmaceutical research find dry rooms more efficient than glove boxes. Walk-in incubators and warm rooms can provide an appropriate environment for culture growth and

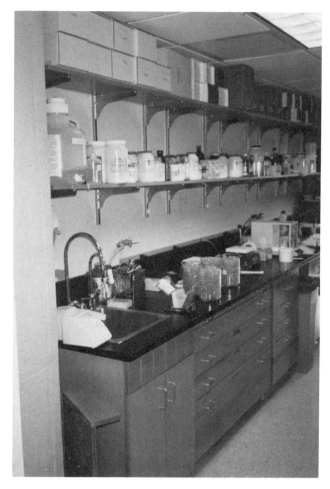

Figure 9.7. Environmental room entrance. Note recording chart, control panel, and safety graphics on the door.

shaker equipment with precise temperature control (±0.15°C) and uniformity (± 0.3°C). High humidity levels can be maintained at prescribed levels as well.

Typically, environmental rooms will need 208 V, 3-phase, 30 amp, 4 wire power at the control panel and 208 V, 3-phase, 25 amp, 3 wire service at remote condensing units. Plug mold, 115 V, single-phase service should be provided in the room at 48" above the insulated floor with outlets spaced at 24 inches on center, except not near the sink. Freestanding and metal wall shelving is often desirable. It is recommended to remotely locate condensers for self-contained environmental rooms in mechanical equipment rooms to reduce heat gain, noise, and vibration. A seven-day chart recorder is recommended, located just outside and adjacent to the insulated room door.

These rooms should be ADA accessible without a curb at the entrance and with open space within the room to permit a 60-inch turning radius for a wheelchair. Unless the room is over 400 ft^2 it should have only one door to help control temperature and humidity fluctuations.

Instrument Rooms

The noisy and high heat generating equipment should be somewhat isolated from the lab work spaces, yet convenient. The minimum and most efficient room width with equipment along facing walls is 10 feet. Equipment rooms for biomedical labs require a variety and density of electric power unmatched in other lab areas. A typical 10 by 20 foot (200 ft^2) equipment room will have 13 power outlets mounted at 48 inches above the floor, with 11 circuits as follows:

- 5 - 115 vac, 20 amp duplex outlets on 3 circuits
- 1 - 208 vac, 20 amp, 3-phase circuit
- 3 - 208 vac, 20 amp, single-phase circuits
- 4 - 208 vac, 30 amp, single-phase circuits

Chemicals, Balance, and Clean Glassware

The chemical storage, balance, and clean glassware storage room or area should be centrally located within the lab. It requires about eight lineal feet of high bench space, a sink, a nearby ice machine for shaved ice, and glass-enclosed wall cabinets for clean glassware. The solid front cabinets for chemicals may need to be lockable. The bench should have a power plug mold strip with two circuits and 115 vac outlets spaced 24 inches on center, except not at the sink location. The ice machine will require a floor drain and a dedicated 115 vac, 20 amp circuit.

Pathology

Research or clinical pathology laboratories involve the use of diseased tissues plus contaminated or infectious materials

Figure 9.8. Chemicals and balance area. Note base cabinet with wide drawers (4') for storing clean glass and plastic columns.

of living organisms. A pathology lab should not be designed in a vacuum. Ideally, the pathologist in charge should be the prime resource to the designer. In many cases, pathology labs are BL-3 facilities. There is the potential for exposure to extremely hazardous bacteria and viruses. Operations include tissue cutting, chemical mixing, mechanical manipulations of biological material, a wide variety of biochemical procedures, microscopy, culturing, sample preparation, and staining. In addition to fume hoods, some bench tops may need continuous exhaust registers at the rear wall of the work surface to carry away harmful solvent fumes.

Pathology labs should not be small spaces (less than 200 ft^2) in order to have adequate air volume for dilution. They need a lockable door with limited access and a glazed observation panel. Refer to requirements for BL-3 labs in Chapter 4.

10

Animal Facilities

Animal Value

The value of animals in biomedical research has increased substantially with the advent of gene transfer technology. Embryo micro-injection and other gene implantation procedures hold great promise in conquering genetic diseases. The value of a single transgenic white mouse could easily exceed $100,000 when the time and effort required to effect a successful gene transfer is considered.

The motive for tender, loving care is obvious. The living conditions for those animals must be kept at or near their ideal environment. Nourishment and drinking water must be available in the proper dosages. Cleanliness, temperature, humidity, fresh air, light and darkness cycles, vibration and sound isolation are all important considerations. The research investigator needs to ensure that his investment is protected. Animals must be kept in a secure environment and isolated from contamination (Figure 10.1).

Although mice and rats are the most common animals used in biological and pharmaceutical research, rabbits are more frequently involved with studies related to diseases of the eye. Swine are a major species used for cardiac research. Other research animals include ferrets, hamsters, guinea pigs, turtles, goats, sheep, calves, cats, canines, and nonhuman primates.

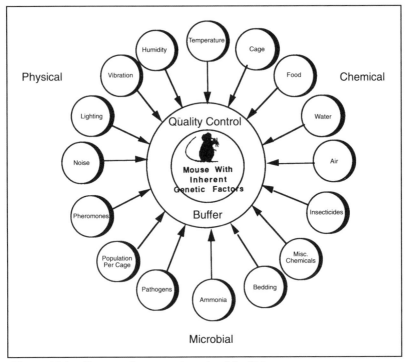

Figure 10.1.   Factors of concern in animal facilities.

Cleanliness

Biomedical research is equally dependent on the quality of the animal supply. They must be viral-free, since any contaminant could lead to misleading data. Some types of scientific investigation require triple-deficient mice (no T, B, or NK cells), nude (thymus-deficient), or germ-free (gnotobiotic) rodents. These species types are very intolerant of change or to conditions beyond their narrow range of acceptable environmental conditions. Separate seven-foot wide access (clean) and return (dirty) corridors should be considered for facilities housing germ-free, triple-deficient and nude mice and rats. This dual corridor concept will minimize room-to-room contamination. In such a facility the air pressure gradient is highest in the access corridor and lowest in the return corridor. To help condition the human occupant, it is a good idea to color code the access corridor differently from the return corridor. Obviously, a dual corridor design is approximately 20% less efficient in usable area and, conversely, 20% more expensive to construct.

Ventilation

Most animal researchers believe that the maintenance of pathogen-free environments requires a minimum of 12-15 clean air changes per hour. Ventilation provides oxygen, removes heat, dilutes gaseous and particulate contaminants and controls pressure relationships among adjacent spaces. Special nude rodents, triple-deficient and gnotobiotic mice can require up to 60 clean air changes per hour with 15 of these changes incorporating conditioned outside fresh air. It is appropriate with these special species to provide an extremely clean environment. Experience has demonstrated that normal viral-free mice and rats (not germ-free) can be maintained satisfactorily in rooms with as few as 12 to 16 fresh air changes per hour. However, odor becomes somewhat objectionable to humans at the lower rate of air change. The recirculated and fresh outside air for animal rooms is directed through high efficiency particulate air (HEPA) filters to limit possible contamination. Prefilters and HEPA filters should be accessible for changing, external to the cage rooms. For all species, it is common practice to install fabric prefilters at the rear face of the return air grills to extend the life of the main fan filters. Generally, it is desirable to develop a laminaire air flow in animal holding rooms with the air supplied at the ceiling and return air grills located low on all walls to ensure adequate coverage. In addition, the air pressure for pathogen-free animal holding rooms should always be positive relative to the corridor, except for quarantine or isolated rooms where there would be a danger of spreading contamination.

Room air should not be recirculated unless gaseous contaminants and particulates have been filtered out. This is especially crucial in rooms housing several hundred animals where the spread of disease could have significant consequences to the colony. Routine maintenance procedures to the recirculating air system are essential. Soiled duct work and fan components can be expensive hidden culprits in animal contamination, resulting in unusable research data. Proper maintenance practices require frequent changes to the fabric prefilters located at the return air grills. Overhead local fan coil units are best located above the corridor ceiling where removable, washable ceiling panels permit convenient access without disruption to the animals housed in the holding or cage rooms.

Temperature and Humidity

Some animals, such as rabbits, have a narrow range of tolerance for both temperature and humidity compared to humans and some other species. For this reason it is good practice to provide separate environmental controls for each animal room. This gives more versatility to long-range uses because animal species can change as the nature of scientific research changes. Proper temperature and humidity conditions are essential for healthy animals since they influence metabolism. Design capability should include a range of 35% to 60% relative humidity. Temperature and humidity readout dials should be located on the corridor side of each animal room. In addition, temperature and humidity sensors located in the return air duct for each animal room can send a signal to a central printout station, such as Security, that is occupied 24 hours a day to alert personnel to system malfunction. The time and date of the aberration should be recorded because this historical data can be important to the researcher. If the budget can accept the cost premium, backup fans should be considered in addition to standby generator power. Systems redundancy can be good insurance for investment protection.

Most human work spaces have less than one fresh air change per hour. The "lungs" or fan power required for research animal facilities are substantially greater, and more expensive to construct and operate, compared to office buildings or hospitals. The building systems for odor and vapor removal are essential to the life and health of the animal colony. Most species of animals require a minimum of 15 air changes per hour. This air is one pass supply. It equates to a complete air volume change every 4 minutes, all day and all night, for the life of the lab. Spaces for housing laboratory animals are temperature and humidity regulated within very close tolerances. Tremendous amounts of energy are consumed by moving large quantities of air and by heating or cooling that air to an acceptable temperature range. Consequently, heating and ventilating systems for animal facilities generally cost between three and six times the initial operating costs for generic wet bench lab space.

Vermin

Generally, pesticides are not used to control cockroaches or other vermin because they can cause harm to the animal colonies themselves. The best designs eliminate breeding and refuge sites. (Floor drains, with their stagnant water traps, should be avoided except where absolutely necessary.) Any cracks and wall openings must be sealed, especially during construction where they are concealed above suspended ceilings.

Room Finishes

Animal room ceilings should have smooth, hard, impervious surfaces uninterrupted with joints or access panels. Typical construction consists of epoxy painted moisture-resistant gypsum board or hard coat plaster with wall-to-ceiling junctions sealed with exterior grade acrylic sealant. Animal room surfaces have to be washed down periodically to disinfect the space. An eight-inch high projecting curb at the wall-to-floor junctions protects the walls from damage from the cage racks. Floor-to-wall junctions should be coved for sanitation.

Concrete block wall surfaces need to be sealed with block filler and coated with a minimum of 8 mil epoxy paint. Moisture-proof, ceiling-mounted fluorescent fixtures must have edges sealed. Weatherproof electrical convenience outlets are necessary in each animal room to provide power for the wet vacuuming equipment. Floor drains are not desirable since they provide a breeding habitat for vermin. Floor finishes must withstand damage from urine. Cleaning incorporates wet vacuuming and sponging with cleaning and disinfection solutions.

All cabinets and countertops in the animal support rooms should be made of nonmagnetic stainless steel with welded seams ground smooth. The joints between the steel and other surfaces should be sealed with a sanitary sealant compound.

Decontamination

Occasionally, it is necessary to decontaminate animal rooms where diseased species have been kept. Formaldehyde gas sealed in a room for one hour under 60% relative humidity conditions has proved effective and practical. Room details must permit positive containment with sealed joints and openings. Paraformaldehyde depolymerized in an electric frying pan set at 232°C is sufficient, provided ten grams of paraformaldehyde for each cubic meter of room volume is used. Dampers in the duct work must be closed and fans turned off during the decontamination period. The dampers should be reopened and fans turned on for at least an hour following the process prior to permitting room occupancy.

Water

A central wall-mounted automatic watering system serving each animal room is important to maintain healthy colonies. Usually, this water is purified beyond human potable standards and kept in constant flow to inhibit bacterial growth. Such a system has many advantages including:

- reduced labor costs
- clean, fresh water supply
- no bottles, stoppers, and sipper tubes to become contaminated
- better cage visibility
- eliminates menial task
- no bottle washer required.

These advantages will be negated, however, if the automatic system is not properly and regularly maintained.

Light and Darkness

Animal room illumination is an important factor in the maintenance of healthy animals. Too much light can cause retinal damage in albino mice and rats. Continuous periods of illumination can affect animal behavior. Like humans, animals expect a regular diurnal lighting cycle. Therefore, it is good practice to provide automatically timed lighting for animal rooms. Locate timer controls adjacent to the animal room entrances on the corridor side. General illumination level of 25-foot candles at floor level is recommended. This provides enough light for routine animal care and housekeeping activities.

Sound

Noisy activities such as refuse disposal and cage washing should be located away from the animal rooms. The best way to minimize disturbances to the animals is to isolate them from the sounds of human activity. Noisy animal species such as swine, dogs, and nonhuman primates should be sound isolated from other species. In general, animals, like humans, are more relaxed in a quiet environment.

Cage Room Doors

Doors to animal rooms should be relatively soundproof (STC 40). A viewing window in the door is desirable, but this window should be covered during the darkness cycle. We typically provide a hinged bulletin board panel at the glazed opening which is normally closed, except during moments needed for room observation. The bulletin board provides a convenient place to post specific temporary data related to animal or current status of housekeeping tasks related to each animal room (Figure 10.2). Doors should not swing into the corridor and they should be at least 42" by 80" to permit regular passage of cage racks. Required door hardware includes heavy duty butts, lock set, closer or spring hinges, armour plate, and automatic drop seal. All edges of the hollow metal door should be closed and sealed to prohibit the possibility of providing a habitat for vermin.

Cage Washing

Providing plenty of spare cages and cage racks is key to efficient animal facility operation. Wire-bottomed rodent

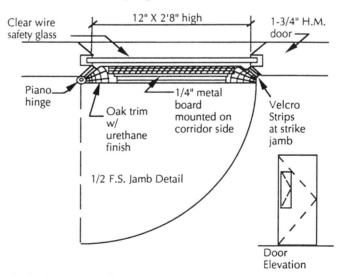

Figure 10.2. Cage room door arrangement.

cages, for example, are washed twice weekly and the portable racks at least monthly. Since the animals must be contained while these cleanings occur, spare cages, racks, and storage space are required. The cage and rack washing area needs an anteroom for stockpiling dirty cages and a post-washing storage area for the clean cages. The use of special cage washing mechanical equipment is highly recommended, with sanitation depending on heat for effectiveness. This cage washing area is served by steam, hot and cold water, and exhaust system to ventilate the steam and humidity. A built-in pit is usually required for the larger cage washing equipment.

Waste

Incineration is one method of waste disposal. However, it is often difficult and expensive to comply with all local, state, and federal regulations regarding the incineration of animal waste products. Waste bagged in plastic and disposed of legally by a qualified contractor is the more popular waste disposal process for most laboratories. A walk-in freezer, located near the necropsy room is used to temporarily store animal tissue and carcasses in plastic bags. Cold storage below 7°C reduces decomposition and putrefaction of biological wastes. Hazardous wastes must be rendered harmless by sterilization, containment, or other appropriate means prior to leaving the animal facility. It is desirable to have an autoclave available in the animal facility for sterilization of bagged waste material.

Inspection and Quarantine Area

A separate, dedicated receiving room for unpacking and animal inspection with adjacent quarantine isolation rooms is considered proper. (It is good practice to have separate isolation rooms for each species.) It is important to have the veterinarian evaluate the health of the newly received animals prior to their introduction to the other similar species to maintain a healthy colony.

Food Storage

The amount of food stored within the laboratory animal facility should be kept to a minimum. It is better to receive frequent deliveries of fresh food than to keep large quantities stored on the premises. Even so, refrigeration may be considered if frequent food deliveries are not practical. In any event, a vermin-proof storage room is required.

Clean and Soiled Bedding Storage

Separate locations for clean and bagged (and autoclaved) soiled bedding storage is needed. These too, need to be kept vermin free.

Operating Suite

Prep rooms with sinks are required for both the surgeons and larger animals that are separate from the operating or procedure rooms. A surgical support room with autoclave and storage space for instruments should be adjacent to the operating room. The anesthetic machines and suture materials are generally stored within the operating room. In addition, a post-operative animal holding room (intensive care) is usually necessary for the larger animals. Sometimes, the animal is kept in a special HEPA filtered cage during the few hours following surgery to keep them free of pathogens and isolated from other animals during the time they are vulnerable. The operating room is fitted out in a similar manner for animals as in a typical hospital with provisions for a scavenging system for exhausting waste gases from anesthesia machines. The room should be kept under positive pressure from the corridor. A dedicated surgery suite is not required for smaller species such as rodents although separate procedure rooms with stainless steel work counters are often required for small animals within the animal facility.

Necropsy Room

A dedicated necropsy room with adjacent freezer for autoclaved and bagged carcasses is usually provided within the laboratory animal facility. The cabinets and countertops and tables are constructed of stainless steel for ease of disinfecting. The necropsy room is kept under negative pressure. Formalin, a combination of formaldehyde with a small amount of methanol, is a clear aqueous solution used to kill infectious bacteria during necropsy procedures. It is also a known carcinogen; therefore, exhaust grills should be provided at the rear of the necropsy work counter to protect the personnel.

Janitor

Janitor's facilities space for wet vacuuming equipment, with floor sink, and shelves is necessary within all animal facilities.

Staff Facilities

A small staff lounge with adjacent toilets and locker room is appropriate for the animal facility workers. It should not have any provisions for food and drink. That should be outside the animal facility enclosure.

Security and Access

The entire facility should be self-contained with sally port type air-lock access points. Coded card access readers are recommended for all entrances to help maintain security from unauthorized persons.

Costs

The veterinarian who will be responsible for supervising the animal farm operations should take an active role advising the architect during the planning phase. Initial construction costs for animal farms are substantially greater than generic wet bench research laboratory space. Ongoing operating costs for animal farms are also significant. The architect should analyze the cost trade-offs carefully with the client in determining the design features incorporated in the facility. As with most laboratory projects, these choices are crucial to the success of the laboratory.

Aquatic Environments

Zebra fish are becoming a popular life form for gene research. They are a plentiful species (available at pet stores throughout the world), inexpensive, fast-growing, and they provide plentiful quantities of eggs for micro-injecting. A typical zebra fish facility needs three spaces: a general room for fish tanks including a large tank washing sink, an adjacent food preparation room, and a separate aquatic laboratory for embryo washing, egg injecting, and where special studies take place (Figure 10.3).

The temperature in all of these spaces must be maintained within the range of tolerance for the fish; for zebra fish that is ideally at 28.5°C (above 31°C and below 25°C is life-threatening to the fish). The fish are maintained in 10-gallon glass tanks which can support about 25 zebra fish. The relative humidity in these rooms is 90% to 95%, so the finishes must withstand the long-term moisture. A vapor barrier enclosure is required with seals at the entry doors. These doors also need panic bar closers for easy opening when carrying a 10-gallon tank. They should also have cypher locks to prevent unauthorized access.

The general fish storage room may contain about 500 fish tanks located on floor to ceiling racks. The floor should be seamless and waterproof with a drain in the event a tank

Figure 10.3. Zebra fish aquaria. Racks of 10-gallon glass fish tanks with circulated aerated water.

spills or is broken. The lighting should be from vaporproof fixtures on a day/night timer with manual override. The water filter pumps should have standby power due to the critical nature of the system for the survival of the fish. Outside air must be continuously pumped into the fish tank water. A large tank washing sink is required in this room (40" x 24" x 24") with adjacent drainboard and flexible spray nozzle. The pumps, filters, and piping for the tank racks should run exposed for easy maintenance. Some lab bench space is desirable for dissecting microscope work in the general fish storage room with gas and vacuum outlets, with deep (18") shelves above for small tanks used for single pair mating.

The food preparation room, located off the general tank room should have a lab bench and sink with deep shelves above. The room will need to accommodate two people. The fish are fed a minimum of twice per day although multiple light feedings are better. Zebra fish love brine shrimp, but they also eat ground dry or moist trout pellets as well as dry flake food available at pet stores. Fish need a variety in their diet for good health and breeding. This room should be under negative pressure with a good exhaust to control odor.

The special aquatic lab needs to have a water filtration system which can be interconnected to the main fish storage room.

11

Clean Environments

Air Quality

An ordinary room in your home has on average 300,000 airborne particles of 0.5 micron or larger for each cubic foot of air. Thus, if it were cleanroom classified, this typical room would be a class 300,000 clean room. A space with air three times cleaner than the normal interior room has only 100,000 (100K) airborne particles for each cubic foot of air. This is the dirtiest cleanroom classification, class 100,000. A room with air ten times cleaner than the class 100K cleanroom would be a class 10K clean room. The lower the classification, the cleaner the room air. A class 10 space is extremely clean, inasmuch as there are no more than ten particles of a size of a half of a micron or larger for each cubic foot of air (353 particles at that size per cubic meter). Cleanrooms or clean work surfaces are rated as class 1, 10, 100, 1K, 10K, or 100K. New requirements for air cleanliness are addressed in the Federal Standard 209E, "Airborne Particulate Cleanliness Classes in Cleanrooms and Clean Zones."

A micron, or micrometer, is one-millionth of a meter. Airborne particles vary in size from 0.001 microns to several hundred microns. Particles larger than five microns in diameter tend to settle fairly quickly. Particles occur in nature as bacteria, pollen, miscellaneous living and dead organisms, seaspray, and windblown dust. Human beings are a prime source of airborne particles. Human skin is constantly regenerating and releasing tiny dead flakes. Respiratory and digestive gas emissions contain particulates. Lint, cosmetics, and dander also contribute to the dirt in the air. Industry generates particles from chemical vapors, combustion processes, and friction from moving equipment parts. With many research and manufacturing processes, airborne particles are a source of contamination and can compromise an experiment or cause product failure.

Leading edge techology has now defined particles equal to or less than .02 micron in diameter as ultrafine particles. In Class 1 and better cleanrooms, devices now can measure the frequency of ultrafines more efficiently than the larger .5 micron particles, since there are so few of the larger particles present in such an environment.

Research often requires an ultra clean air environment at the work surface and extremely low background levels of impurities measured in parts per billion (ppb). Some research activities need the entire room air clean-classified. Airborne particles are viewed as a source of contamination. Obviously, the cost to achieve cleanliness increases as the level of classification decreases, and larger clean space volumes are more expensive to build and maintain than smaller volumes. Cleanroom design requires a level of engineering sophistication much higher than more ordinary building systems. Cleanroom performance depends on controlling the particulate concentration and dispersion, temperature, humidity, vibration, noise, air flow pattern, illumination, pressurization and construction of surface materials. The objective of a good cleanroom design is to control these potential variables within acceptable parameters while maintaining reasonable initial and operating costs.

There are four basic concepts to follow in designing a clean facility:

- Pressurize the room at all times.
- Provide clean air supply to the room.
- Dilute the contaminated air.
- Keep air turbulence to a minimum.

Pressurization

A clean space should be maintained at a higher static pressure than its surroundings to prevent infiltration by wind or other effects. All cleanrooms, regardless of their classification, must be under pressure with respect to the natural atmosphere. There are often multiple cleanrooms with differing requirements for contamination control. Differential pressures must be maintained between cleanrooms sufficient to assure airflow outward progressively from the cleanest spaces to the least clean at all times, even when a connecting door is suddenly opened. Any sudden pressure change can shock the dust free from filters and cause a sudden cloud of dirty air to penetrate the protected environment. It is important to keep air flow variations to a minimum. Air locks or vestibules that are maintained at slightly reduced pressure are incorporated in the entrance design.

Static pressure regulators can maintain the desired room pressures by operating dampers, fan inlet vane controls, vane axial fans, controllable-pitch-in-motion controls or a combination of these to vary the ratio of supply air to make-up air or exhaust air. Variation in air flows should be minimized while maintaining control over room pressures.

Air Filtration

Air supplying the room must be clean. This is achieved with the use of High Efficiency Particulate Air (HEPA) filters or Ultra Low Penetrating Air (ULPA) filters. HEPA filters are 99.97% to 99.997% efficient in removing 0.3 micron particles and larger. ULPA filters are even better. They are 99.9997% efficient in removing 0.12 micron particles. Both of these super air filters use glass fiber paper constucted in a pleated format. These filters are normally two by four feet and can be directly connected to a supply air duct. They are also available with individual fans that can direct air through the filter from an air supply plenum.

HEPA filters were developed in 1962 to meet the growing demand for the clean environmental conditions needed to produce computer chips. The HEPA filter is a disposable dry-type filter, constructed of boron silicate microfibers cast into a thin sheet, similar to a piece of paper. This filter media is pleated to increase its surface area. Corrugated metal separators are placed between the pleats to permit the air to penetrate deep within the pleat. The HEPA filter retains airborne particles and microorganisms while permitting gases to pass through freely. Five distinct mechanisms allow the HEPA filter to function: electrostatic attraction, interception, sedimentation, diffusion, and inertial impaction.

HEPA filter life varies a great deal with the hours of operation, amount of particulates (dirt) in the laboratory, and the nature of the work being performed. Typically, however, HEPA filters do not need to be replaced more frequently than once every 3 to 5 years.

Air Quantity

Enough clean air must be provided to dilute the contamination created by the occupants and the work processes. It is necessary to increase the air flow rates as the cleanliness classification decreases. A class 100 cleanroom is much windier than a class 10K room. Typical air flow rates for successful cleanrooms are as follows:

CFM/SF	CLASS
80–100	100
25–30	1,000
8–10	10,000
3–5	100,000

It is sometimes necessary to increase the flow rates for the lower classes of rooms to compensate for abnormally high levels of process contamination.

Air Flow

Air turbulence within the room must be minimized because this tends to mix the dirty and clean air. Avoidance of air turbulence is achieved by spreading the air supply evenly over one surface of the room, with the opposite surface as the air return. This creates a laminar flow of air, or parallel air streams at the same speed. This can be done by having the entire ceiling made of HEPA filters supplying air and the floor an open grating into a return air plenum. For relatively narrow cleanrooms, the return air grills can be located low along the side walls. This design does not produce true laminar flow; rather, it can be termed unidirectional air flow. Such a design tends to compromise the air quality at the midpoint between the side walls in the lower area of the space.

Most cleanrooms do not have laminar air flow, but unidirectional air flow. It is characterized by air flowing in a single pass in a single direction through a clean space or zone with generally parallel streamlines. Ideally, these flow streamlines are uninterrupted and, although people and equipment in the air stream do distort the streamlines, a state of constant velocity is approximated. Most particles that encounter an obstruction in the laminar airflow strike the obstruction and continue around it as the airstream re-establishes downstream. Air turbulence is strongly influenced by the air supply and return configurations, movement within the room by people or robots, and by equipment placement within the room.

Laminar airflow or unidirectional airflow is required for cleanrooms below Class 1,000. It is possible to achieve a Class 1,000 (1K) level of cleanliness with a multidirectional airflow air pattern. In order to maintain such a cleanliness level with multidirectional airstreams presupposes that the major space contaminator is from the make-up air (external source) and that such contamination is removed at the air handler or ductwork filter housing or through HEPA filtered supply devices. When internally generated particles are of primary concern, clean workstations such as safety cabinets, hoods, or glove boxes, must be provided in the clean space.

Room Construction

A cleanroom is conceived as a room within a room (Figure 11.1). The inner room is where the work occurs and is occu-

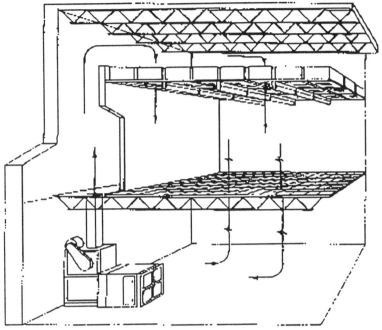

Figure 11.1.   Laminar flow cleanroom with HEPA filter ceiling supply and continuous floor grill with recirculating fan located on floor below clean room.

pied by people. The outer room is air plenum, duct, and sometimes fan equipment space. It usually requires as much or more volume as the occupied space. The outer room is pressurized and sealed from its surrounding space. The inner room is sealed as well except at the designated air supply and return openings. The architect and the mechanical engineer must have close collaboration to create a successful cleanroom. Usually, the walls of the inner room function as return air ducts while the ceilings are almost entirely air diffusers. The rooms themselves must be thought of as vessels to maintain positive pressure with respect to the surrounding spaces. The list of design concerns includes:

- class of room cleanliness
- temperature and humidity requirements for the tasks to be performed in the space

- availability and quantity of electric service
- availability of reliable, 24 hour, year around heating source
- availability of reliable, 24 hour, year around cooling source
- amount of exhaust required
- structural integrity of enclosure to maintain positive pressure
- available space for mechanical equipment and ductwork
- structural support for mechanical equipment and maintenance personnel
- duct static pressure analysis
- thermal insulation requirements
- provisions for air balancing and adjustments
- need for backup equipment
- adequate illumination at the working area
- background noise level requirements
- vibration sensitivities of work tasks.

Temperature

Stable temperatures are required in a cleanroom to maintain stable conditions for materials, experiments, instruments, and staff comfort. The heat load from lighting is high but stable; personnel loads can vary, as can the heat generated by any process or equipment use. Large amounts of supply air in cleanrooms diffuse the internal heat gain so that the temperature differential between entering room air temperature and exiting room air temperature is low. However, the locations for heat producers and air supply patterns need to be analyzed to determine the resulting temperature gradients. Large cleanrooms may require multiple zones of temperature control because of the differing cooling requirements of localized areas.

Humidity

Humidity control is necessary for a number of reasons; to wit: prevent corrosion or oxidation, prevent condensation, reduce static electricity, provide personnel comfort, compensate for hydroscopic materials, prevent product contamination, and to control microbial growth.

Humidity control is affected more by external influences, such as changes in weather, than by the variations in moisture generation within a cleanroom. If there are processes involving evaporation in a cleanroom, they should take place within a hood. It is not unusual for precision manufacturing processes and certain experiments to require low humidity conditions. Under these conditions, precautions must be taken to control static electricity at equipment with ionization grids and grounding straps.

Make-Up, Exhaust Air

Activities conducted in a cleanroom often require the use of a fume hood with its exhaust. These work stations use the cleanroom air as their source of make-up. Therefore, the cleanroom air supply volume must be increased by the volume of the exhaust. It is important to orient the fume hoods in a cleanroom to maintain the unidirectional airflow within the space.

Noise, Vibration

The high air velocities, contamination control filters, and fans necessary to maintain cleanliness levels make it very difficult to control background noise in cleanrooms. The noise and vibration criteria required within the cleanroom should be established prior to commencing the schematic design. Electronic microscopes and other ultrasensitive equipment may have certain sound or vibration frequency ranges that are critical to avoid. In applications where low sound levels are of utmost importance, fans and filters may have to be remote and mounted with dampening devices to reduce the equipment operating sound pressure levels. These measures can greatly increase the initial constuction costs.

Odor Removal

Ventilation, adsorption, odor modification, and chemical reaction are some of the tools used in HVAC systems to remove odors. For water soluble odors, such as ammonia, water sprays, packed scrubbers, or wet cooling coils in the air stream will remove odorus gases or vapors. Exhaust air containing objectionable gaseous odors, irritants, or particulates that obscure vision and toxic matter should be diluted and replaced with clean outside air. "The solution to pollution is dilution."

Adsorption is the adhesion of a gas or vapor on an activated solid substrate. Activated carbons are made from materials such as coconut shells, coal, peat, and petroleum residues, which are heated in reducing atmospheres to produce high porous materials with enormous surface areas per unit volume. The activated material is placed in filter frames through which air is passed. Removal of impurities varies according to the physical properties of the impurities, thickness of the filter material, the air velocity, and the pore size distribution of the activated surface. When activated charcoal, an efficient odor removal agent, has adsorbed its full capacity, it is removed and replaced with fresh material. Some activated materials can be regenerated for reuse by the supplier.

Many odors can be destroyed by oxidation. While oxidizing gases such as ozone and chlorine can oxidize odors in water, concentrations required for air deodorization would be toxic to humans. The removal of water-insoluble odors and certain types of collected odorants is increasingly difficult. A final option to removal or destruction of objectionable odorants is to introduce other chemicals. They can either mask the smell to make it more acceptable, or chemically neutralize the malodor to an acceptable intensity level.

Room Finishes

In general, the finish surfaces for a cleanroom should be smooth, nonshedding, as monolithic as possible, cleanable, chip resistant, with minimum seams, joints, and no moldings or crevices. All penetrations for pipes, ducts, conduits, etc., should be fully sealed and gasketed. Door frames, vision panels, switches, clocks, lockers and other appurtenances should be flush mounted where possible or have sloped tops to minimize air turbulence.

Where the floor surface is not functioning as a return air grill it should have an integral base flashed up the wall about 4" to 6". Seamless sheet vinyl, epoxy, or polyester coating are satisfactory clean room flooring materials.

Cleanroom wall surfaces are often epoxy-coated drywall, baked enamel polyester or laminated plastic with minimum projections that could cause air turbulence. The wall finish may also serve as part of the vapor barrier for spaces that require humidity control.

Ceilings in Class 100 or cleaner rooms are usually composed entirely of HEPA filters. Otherwise, ceilings are plaster or drywall covered with epoxy or polyester coating, or plastic finished acoustic panels. Where a ceiling grid is used to support HEPA filters or ceiling panels, a continuous closed cell gasket or sealant is placed between the tee grid member and ceiling panel to provide an airtight joint. Ceiling lights are teardrop shaped, single lamp fixtures mounted between and below filter joints or flush mounted and sealed.

Operating Procedures

Equipment and materials must be thoroughly cleaned before entry. This even applies to workmens' tools during the final stages of initial construction operations. All work personnel should be trained in proper cleanroom procedures prior to entry.

Pencils and erasers are not permitted. Nonshedding paper and ballpoint pens are used. Disposable nonparticulating fabric "bunny" suits, caps, goggles, face masks, and booties are worn by the occupants. To reduce contamination, gowning takes place in a special area at the air lock entrance to the clean room suite. Room air pressure differential must always be highest in the cleanest area, and descend to above atmospheric conditions at the air lock.

Room air pressure is reduced as cleanliness requirements become less stringent. Cosmetics may not be worn by clean room workers. Human hair should not be exposed, to avoid transfer of skin oils and dander particles. Work is handled with gloved hands, finger cots, tweezers, vacuum wands and the like, to minimize contact with workers.

Any particulate-producing operations such as grinding, welding, and soldering in a cleanroom must be shielded and independently exhausted. Sealed containers are used for material transfer and storage. The most sophisticated cleanroom design can be easily compromised by naive, untrained personnel using improper operating procedures.

Clean Benches

The laminar flow clean bench, when properly used and maintained, provides product protection from ambient contamination. It ensures that the product is exposed only to HEPA filtered, particulate-free air. Applications include electronic assembly and testing, plant tissue culture, media preparation, and syringe filing.

12

Lab Equipment

Laboratory environments and services are designed to satisfy the equipment requirements as much as the human needs. This equipment is often highly sophisticated. Measurements of parts per billion (ppb) are commonplace. Super-computers, lasers, high energy magnets, radiation, X-ray, temperature extremes, ultra-clean and pure fluids are all encountered in the modern research laboratory setting.

To comprehend in detail the workings of all complex laboratory equipment is beyond the comprehension of the architect. A cursory knowledge of this equipment is important to communicate and understand the needs of the researcher.

During the design development work, the architect will work closely with the laboratory user to understand and document the exact needs of each piece of laboratory equipment. These include: physical size, required clearances, hazardous conditions, weight, vibration tolerance, electrical characteristics, emergency or standby power, heat generation in Btu, gas and vacuum connections, lighting environment for proper operation, radiation shielding, pure or cooling water connections, drains, steam, compressed air, etc. In addition, the anticipated hours of usage per day, or diversity of each piece of equipment should be predicted, if possible.

Special Equipment

Listed below are some of the more common equipment items found in a biomedical laboratory along with very general descriptions of their requirements. Specific equipment, environmental and service needs should always be confirmed with the manufacturer, as technology is constantly evolving. Beware; required services may differ among manufacturers of the same generic equipment item.

Autoclave/Sterilizers and Stills (Figure 12.1): Uses steam under pressure as sterilizing agent.

Approx. footprint size: 40"W x 50" to 76"D x 75"H; plus, allow 18" clear (Autoclaves are usually specified in terms of chamber dimensions)

Services required: cold water, *steam supply, steam return (vented), electric power, drain, indirect exhaust

Approximate heat output: 12,000 Btu - May be freestanding or recessed.

* May be "house supplied" or ordered with an electric steam generator.

Amino Acid Analyzer: Special benchtop computer

Approx. size: 31"W x 21"D x 17"H

Services required: dedicated electric power (isolated ground), high purity argon gas, pure water, exhaust duct

Biological Safety Cabinet: Negative pressure hood with unidirectional HEPA filtered room air. Basic item for tissue culture work.

Approx. size: 56" to 80"W x 34"D x 88"H

Services required: electric power, indirect exhaust, natural gas, vacuum

Heat output: 1,500 Btu

Cage and Bottle Washer, Small: Used in care of small laboratory animals.

Approx. size: 71"W x 139"D* x 75"H; (*with both doors open)

Service required: electric power, steam, hot and cold water*, condensate return, compressed air, drain, exhaust vent;

(*Cages must have 180°F wash temperature via integral 140° to 180°F booster heater with "stop" programming. The water must be tempered from 180°F to 140°F before entering the drain.)

Cage and Rack Washer, Large: Used in care of larger laboratory animals.

Approx. size: 63"W x 94"D* x 127"H; (*with both doors open)

13" deep pit required

Figure 12.1.   Autoclave, used to sterilize material.

Service required: electric power, steam, hot and cold water*, condensate return, compressed air, drain, exhaust vent; (* Cages must have 180°F wash temperature via integral 140° to 180°F. booster heater with "stop" programming. The water must be tempered from 180°F to 140°F before entering the drain.)

Centrifuge and Ultra Centrifuge: Revolving rotor about vertical axis under controlled temperature. Uses centrifugal force with vacuum to separate solutes from solvents. Usually these are placed in a separate room because of heat, noise, and biohazard potential (Figure 12.2).

Approx. size: 28" to 32"W x 33" to 39"D x 36" to 63"H

Service required: electric power

Approx. heat output: High speed centrifuge - 14,000 Btu, 80% diversity

Low speed centrifuge - 10,000 Btu, 80% diversity

Ultra-centrifuge - 7,000 Btu, 80% diversity

Figure 12.2. Centrifuges in an open instrument with "unistrut" mounted power mold.

Cesium Irradiator: Container for biological or botanical irradiation and sterilizations. (Keep in a separate restricted area.)
Approx. size: 32" Diameter x 60" to 78"H
Weight: 2,000-6,000 lb.
Sevices required: Electric power, compressed air

Chromatography Drying Oven:
Approx. size: 34"W x 33"D x 45"H
Service required: electric power, direct exhaust vent
Approximate heat output: 11,400 Btu
May be explosion proof

Horizontal Laminar Flow Cabinet: Positive pressure cabinet with unidirectional HEPA filtered room air.
Approx. size: 38" to 98"W x 36"D x 64" to 70" H
Service required: electric power; possibly gas and vacuum.
Approx. 500 Btu/LF of access opening width

Desiccator Cabinet: Enclosure for storing laboratory materials under low humidity conditions
Size: varies considerably
Service required: electric power

DNA and RNA Synthesizer: Tabletop unit located adjacent to heating bath
Approx. size: 31"W x 21"D x 27"H
Services required: dedicated electric power (isolated ground), high purity argon gas, direct vent to fume hood

Electron Microscope: (Nonoptical) Special environment required: dimming lights, vibration-free, cable troughs, humidity control, magnetic shielding, adjacent dark room, isolated (10') from other EMs, darkroom light, dust-free atmosphere.
Approx. room size: 110"W x 120"D x 100"H
Services required: dedicated electric power, compressed air, chilled water (or heat exchanger), drain, dry nitrogen, liquid nitrogen

Electrophoresis Equipment: Instrument separation procedures sometimes involving high voltages such as power supplies, cells, gel dryers, blotters, density gradient fractionators and peristaltic pumps. Shelf or countertop mounted.

Flow Cytometer: Countertop laser for cell analysis
  Approx. size: 86"W x 25"D x 37"H (includes req. clearances)
  Services required: dedicated electric power (isolated ground),
  cold water, drain, compressed air

Freezer - Undercounter Type: (–10 to –20°C)
  Approx. size: 24"W x 23"to 26"D x 34"to 35"H
  Service required: electric power, emergency power
  Front intake required for avoiding overheating of refrigera-
  tion system.

Gamma Irradiator: Container for biological or botanical irradia-
  tion and sterilizations. (Keep in restricted area.)
  Approx. size: 37"W x 49"D x 58"H
  Weight: 6,300 lb
  Sevices required: electric power, compressed air

Gas Chromatography System: Gas analyzer with oven
  Countertop computer components
  Approx. size: 54"W x 26"D x 15" to 31"H
  Service required: electric power

Gas Sterilizer: Uses ethylene oxide gas (flammable, toxic) as
  sterilizer.
  Approx. size: 77"W x 71"D x 70"H
  Services required: well ventilated space (10 air
  changes/hour)compressed air, dedicated direct vent to
  outside, electric power, direct exhaust

Glassware Dryer/Sterilizer: Generally located in a dedicated
  glasswashing room.
  Approx. size: 39"W x 37"D x 77"H
  Service required: electric power, exhaust vent

Glassware Washer: Automatic washer and detergent which
  clean and dry laboratory glassware. Generally located in a
  dedicated glasswashing room.
  Approx. size: 69"W x 39"D x 86"H (includes req. clearances)
  Services required: steam or electric heated, drain, compressed
  air, deionized water.
  Approximate heat output: 12,000 Btu

Glassware Dryer: Generally located in a dedicated glasswashing
  room.
  Approx. size: 69"W x 39"D x 86"H (includes req. clearances)

Services required: electric power
Approximate heat output: 12,000 Btu

Glove Box: Enclosure that provides a physical barrier between
the product and the worker or environment. This is found
in BSL-3 and BSL-4 labs.
Approx. size: 30" to 48"W x 30"D x 30"H
Services required: electric power, nitrogen, vacuum optional

HEPA Filtered Animal Rack: (May be positive or negative flow)
Approx. size: 80"W x 38"D x 74"H
Service required: electric power, emergency power

Histology/Pathology Cabinet: Enclosure which rids the labora-
tory of noxious solvent vapors, including xylene and
formaldehyde.
Approx. size: 48"W x 28"D x 48"H
Service required: electric power
Charcoal filters (need occasional changing)

HPLC, Radio-Chromatography Detector System: Detector for
high energy beta (P-32) or low energy gamma emitters (I-
125) counting.
Countertop computer components
Service required: electric power

Ice Maker: (Shaved, not cubed ice)
Approx. size: 39"W x 30"D x 41"H
Service required: electric power, drain, potable water with
good filter system

Incubator, Large Capacity:
Approx. size: 41"W x 35"D x 90"H
Service required: electric power, emergency power, water,
drain
Heat output: 4,100 Btu, 60% diversity

Incubator, Double Upright: Typically used in tissue culture
rooms
Approx. size: 25"W x 24"D x 86"H
Service required: electric power, CO_2
Heat output: 700 Btu

Incubator/Shaker:
 Approx. size: 41"W x 29"D x 66"H
 Service required: electric power, emergency power

Laminar Flow Cabinet: A Class II Biological Safety Cabinet offers personnel, product, and environmental protection
 against low to moderate risk pathogens. See Biological
 Safety Cabinet
 Services: Compressed air, vacuum, and sometimes a dedicated exhaust, emergency power.

Lyophilizer (Freeze-Dryer), Large: Removes moisture from a
 sample under vacuum to preserve it
 Approx. size: 45"W x 29"D x 57"H
 Service required: electric power

Lyophilizer (Freeze-Dryer), Portable: (on casters). Removes
 moisture from a sample under vacuum to preserve it
 Approx. size: 25"W x 24"D x 36"H
 Service required: electric power

Microtome/Cryostat: Used to make thin slices of frozen speci-
 mens for analysis and study. Locate away from path of air
 movement.
 Approx. size: 28"W x 31"D x 31"H
 Services required: electric power

Protein/Peptide Sequencer: Mounted on freestanding table, relative constant temperature environment required, creates
 hazardous wastes
 Approx. size: 48"W x 30"D x 36"H
 Services required: dedicated electric power (isolated ground),
 prepurified nitrogen, continuous dedicated direct exhaust
 vent

Peptide Synthesizer:
 Approx. size: 59"W x 33"D x 61"H (includes req. clearances)
 Services required: dedicated electric power (isolated ground),
 prepurified nitrogen, continuous direct exhaust vent

Refrigerator - Undercounter Type: (4°C)
 Approx. size: 24"W x 23" to 26"D x 34" to 35"H
 Service required: electric power, emergency power
 Front intake required to avoid overheating of refrigeration
 system.

Refrigerator - Upright Type: (4°C)
Approx. size: 32" to 78"W x 31" to 35"D x 80" to 84"H
Service required: electric power, possible condensate drain

Scintillation/Gamma Counter:
Approx. size:34"W x 31"D x 26"H
Service required: electric power
Heat output: 4,000 Btu, 60% diversity

Steam Generator: (Often used with autoclaves, glasswashers and cage washers)
Approx. size: 20" to 28"W x 30" to 45"D x 28" to 36"H
Service required: electric power, water, drain

Tunnel Cage Washer: Used in care of laboratory animals
Approx. size: 444"W x 65"D x 75"H
Service required: electric power, steam, hot water, condensate return, drain, compressed air, exhaust vent

Ultra Low Temperature Freezer - Chest Type: (–20° to –135°C)
Generally located in separate equipment room because of noise and high heat output.
Approx. size: 29" to 96"W x 27" to 29"D x 41"H
Service required: electric power, emergency power
Approximate heat output: 3,500 Btu
Frequently tied into central alarm system

Ultra Low Temperature Freezer - Upright Type: (–20° to –85°C)
Generally located in separate equipment room because of noise and high heat output (Figure 12.3).
Approx. size: 34"W x 29" to 35"D x 80"H
Service required: electric power, emergency power
Approximate heat output: 3,000 Btu
Frequently tied into central alarm system

Ultra-Pure Water Unit: Wall-mounted above sink. Water purification equipment using reverse osmosis and deionization technologies packaged to conveniently deliver high purity water.
Approx. size: 37"W x 7"D x 25"H
Services required: R.O.D.I. water, electric power, drain

Figure 12.3.   Low temperature freezers; larger, noisier and with much more heat output than the domestic variety.

Vacuum Evaporator:
  Approx. size: 30"W x 18"D x 52"H
  Services required: electric power, water

X-O Mat Processor: X-ray film developer, counter-mounted, darkroom with through wall opening
  Approx. size: 49"W x 27"D x 20"H
  Services required: electric power, water, drain, exhaust duct, silver recovery system

During the design development phase for a new laboratory it is recommended that the lab user and the architect develop an equipment description matrix. Most of the information on this matrix is needed by the mechanical and electrical engineers for designing the building services. The contractor, or builder, will also need much of this information in order to coordinate the construction work.

Biological Safety Cabinets

The term "biological safety cabinet" (BSC) is used to describe an aerosol containment device equiped with HEPA filter(s) that has been designed to provide protection to personnel, or to both products and personnel from biohazardous materials. The research community will often refer to the BSC as a "hood," but it is not technically considered a hood. These BSCs come in several varieties and the type selected is the responsibility of the principal investigator (PI) or the industrial health (IH) officer, not the architect. Biological safety cabinets should be certified as conforming to National Sanitation Foundation (NSF) regulations by an independent certifier before initial use, anytime the cabinet is moved, following any accident or spill within the cabinet, and at regular annual intervals.

The Class I biological safety cabinet provides personnel and room environmental protection only and is suitable for work with agents that require Biosafety Level 1, 2, or 3 containment (Figure 12.4). It is similar in operation to a chemical fume hood, except that a Class I cabinet has a HEPA air filter at the exhaust outlet, and does not have to be connected to an exhaust duct. Airflow is away from the operator and with a minimum inflow face velocity of 75 feet per minute (fpm). The Class I cabinet offers no protection to the product within the hood, and therefore is very limited in its application.

A Class II BSC or laminar flow cabinet is defined as a ventilated cabinet for personnel, product, and environmental protection. It has an open front with inward air flow of usually 100 fpm for operator protection, downward HEPA filtered laminar air flow for product protection, and a HEPA filtered exhaust for environmental protection. It is suitable for work with agents classified as having low to moderate risk. Class II cabinets are available in several different configurations: Type A, Type B1, Type B2 and Type B3 (or A/B3 convertible).

The Class II, Type A biological safety cabinet (1) maintains a minimum calculated inflow velocity of 75 fpm through the work area access opening; (2) has HEPA filtered

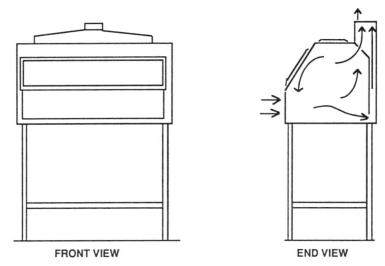

Figure 12.4. A Class I biological safety cabinet.

downflow air from a common plenum (i.e., plenum from which approximately 30% of the air is exhausted from the cabinet and the remainder, 70%, supplied to the work area); (3) may exhaust HEPA filtered air back into the laboratory; and (4) may have positive pressure contaminated ducts and plenums within the cabinet. In fact, Class II, Type A cabinets seldom have a dedicated direct exhaust connection to the outside. It is suitable for work with low to moderate risk biological agents in the absence of volatile radionuclides and volatile toxic chemicals (Figure 12.5).

The Class II, Type B1 biological safety cabinet (1) maintains a minimum average inflow velocity of 100 fpm through the work area access opening; (2) has HEPA filtered downflow air composed largely of uncontaminated recirculated inflow room air; (3) exhausts most of the contaminated downflow air through a dedicated duct exhausted to atmosphere after passing through a HEPA filter; and (4) has all biologically contaminated ducts and plenums under negative pressure, or surrounded by negative pressure ducts and plenums (Figure 12.6). The dedicated and sealed exhaust system should have a remote fan and alarm system for fan

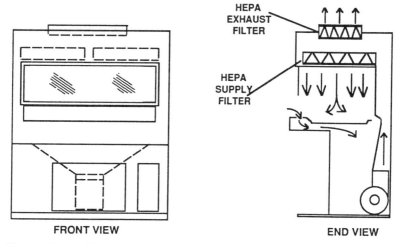

Figure 12.5. A Class II, Type A biological safety cabinet.

failure. The cabinet recirculates approximately 30% of air into the room and exhausts approximately 70% of air to the outside. The sizing of the remote exhaust fan is critical and should be coordinated with the specific cabinet manufacturer. The Class II, Type B1 cabinet is suitable for work with

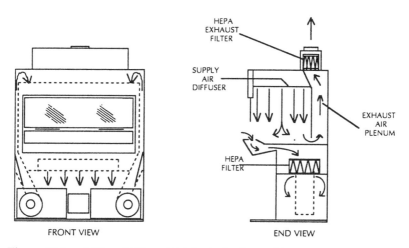

Figure 12.6.   A Class II, Type B biological safety cabinet.

low to moderate risk biological agents, and with agents treated with only a few micrograms of toxic chemicals and trace amounts of radionuclides required as an adjunct to microbiological studies if work is done in the direct exhausted portion of the cabinet.

The Class II, Type B2 biological safety cabinet is sometimes referred to as "total exhaust" BSC. It (1) maintains a minimum average inflow velocity of 100 fpm through the work area access opening; (2) has HEPA filtered downflow air drawn from outside the cabinet; (3) exhausts all inflow and downflow air to the atmosphere after HEPA filtration without recirculation within the cabinet or return to laboratory room air; and (4) has all contaminated ducts and plenums under negative pressure, or surrounded by directly exhausted (nonrecirculated through the work area) negative pressure ducts and plenums. The Class II, Type B2 cabinet is suitable for work with low to moderate risk biological agents and with agents treated with toxic chemicals and radionuclides required as an adjunct to microbiological studies.

The Class II, Type B3 or A/B3 (convertible) biological safety cabinet (1) maintains a minimum average inflow velocity of 100 fpm through the work access opening; (2) has HEPA filtered downflow air that is a portion of the mixed downflow and inflow air from a common exhaust plenum; (3) discharges all exhaust air (30% of total air flow) to the outdoor atmosphere after HEPA filtration; and (4) has all biological contaminated ducts and plenums under negative pressure, or surrounded by negative pressure ducts and plenums. The Class II, Type B3 cabinet is suitable for work with low to moderate risk biological agents treated with minute quantities of toxic chemicals and trace quantities of radionuclides that will not interfere with the work if recirculated in the downflow air.

These different types of Class II biological safety cabinets were developed to meet the needs of research. The National Sanitation Foundation has developed a Standard No. 49 which establishes minimum materials, design, construction, and performance requirements for these Class II laminar flow cabinets. The NSF also makes annual unannounced audits of the cabinet manufacturer's records and cabinet installations. In order to be certified by the NSF, the Class II cabinets must pass three separate tests; to wit: the Personnel Protection Test, which measures the amount of bacterial spores that escape from the cabinet's work area into the environment; the Product Production Test, which measures the amount of bacterial spores entering the work area from the outside environment; and the Cross Contamination Test, which measures how far bacterial spores generated within the cabinet drift across the work area.

The Class III cabinet provides complete (100%) containment (Figure 12.7). It is nominally called a glove box. It is a totally enclosed, ventilated cabinet of gas-tight construction. Operations in the cabinet are conducted through attached rubber gloves. The cabinet is maintained under negative air pressure of at least 0.5 inches (12.7 mm) water gauge. Supply air is drawn into the cabinet through HEPA filters. The exhaust air is treated by double HEPA filtration, or by HEPA filtration and incineration.

Researchers, not the architect, should select the equipment to be used in the laboratory. This is especially true for Class II laminar flow biological safety cabinets. Safety is the primary consideration and it is dependent on the type of protection required—product protection only, personnel, and

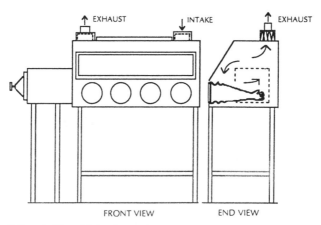

Figure 12.7.   A Class III biological safety cabinet.

environmental protection only, or product, personnel, and environmental protection. All Class II NSF certified cabinets are suitable for Biosafety Levels 1, 2, and 3 containment. No one type offers superior aerosol containment over the others.

Many options and accessories are available which customize a BSC to the user's needs (Figure 12.8). Cabinets come in various widths. The actual dimensions vary with the manufacturer. The most common sizes of biological safety cabinets are 3', 4', and 6' wide. The most popular width is nominally 4' (actual dimension: 4'-7"). Service valves for pure water, vacuum, air, or a variety of gases can be furnished with the BSC; however, the use of flammable gases or solvents should be avoided. Supplementary electrical convenience outlets with ground fault interrupter circuit are often part of the cabinet assembly.

Ultraviolet (UV) lamps may be included as an option in a BSC as an aid in decontamination of the work area. The wavelength of UV light is disruptive to DNA molecules, resulting in a broad spectrum disinfection. While UV light is effective when it strikes a microbial cell directly, it is ineffective if the cell is protected by dust or organic matter. UV irradiation of the work area should only be used as a supplementary method of maintaining the disinfected status of the cabinet. It should not be relied on alone to disinfect a contaminated work area. UV light is irritating to the eyes, and the UV lamp should be turned off while actively working in the cabinet. Most scientists consider UV not to be effective in Class II cabinets where there is so much air movement.

Similar to chemical fume hoods, biological safety cabinets should be located away from traffic patterns, air diffusers and registers, and other hoods that could disrupt air flow across the cabinet face. They should never be installed in a space with windows that open to the exterior. It is important to have at least 6" clear above the cabinet exhaust outlet and any overhead ceiling or obstruction, and where possible 12" clearance is desirable at the rear and both sides of the unit. Biohazard cabinets should be stationary and not mounted on wheels.

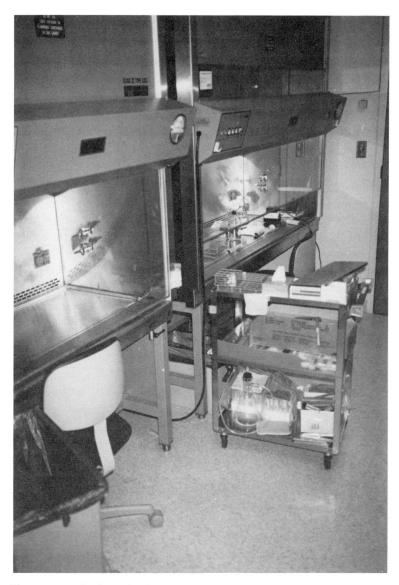

Figure 12.8.   Biological safety cabinets.

13

Shielding

Most research facilities require some form of shielding to protect personnel, equipment and/or information.

Electromagnetic Interference

There is a growing proliferation of sensitive, highly complex electronic and telecommunications equipment being used in our research labs today. During World War II the military and the Federal Communications Commission began to realize the importance of shielding radio frequency interference (RFI). Since then, the general public now routinely uses electronic equipment containing thousands of low voltage integrated circuits which are highly susceptible to electromagnetic interference (EMI). Sophisticated microwave technology has added to the man-made EMI background over an expanded frequency spectrum.The EMI spectrum includes RFI, microwaves, and radar.

The purpose of electromagnetic shielding is either:

- To protect sensitive electronic equipment, generally computers, from high level sources of EMI (computers located near airports are an example). EMI can corrupt computer data; or
- To protect confidential or proprietary information being processed on computers from interception by unauthorized persons. (It is actually possible to detect and analyze the electromagnetic waves emanating from computer equipment.)

Electromagnetic shielding is required in certain installations to create a protective barrier around sensitive computerized equipment to prohibit the adverse effects of electromagnetic interferences (EMI) in the form of radar, radio frequency, and microwave "noise" from entering the space housing the computers. Without adequate shielding, data being processed can be lost or significantly degraded.

The user of the facility must analyze any potential problem which might include a site or equipment survey. Government installations have guidelines and mandated regulations to follow regarding EMI barrier construction. Usually, shielding requirements are specified as a ratio of the electromagnetic field intensity on one side of the barrier to the field intensity on the other side. The frequency range of interest is generally between one kilohertz (kHz) and ten gigahertz (GHz).

There are two types of EMI. The first type is called conducted interference. This is any undesirable electrical or electronic signal which is conducted over power lines or interconnecting signal cables. The second type is radiated interference. This is any undesirable signal which is transmitted through space either intentionally or unintentionally from transmitters, power cables, electrical equipment, and natural sources such as lightning or solar flares.

Electromagnetic shielding is accomplished by a metallic barrier. The specific material is selected based upon the frequency range, ratio of differential energy to be shielded, characteristics of the metal material such as conductivity and permeability and, of course, cost. In some cases, it is possible to make use of the earth for completing a shield in lieu of a basement floor. Shielding may be installed outside the enclosure, as an integral part of the enclosure, or within the enclosure, depending upon the building design, materials selected, shielding requirements, and cost.

Architectural electromagnetic shielding involves surrounding the space to be protected with a continuous, electrical conductive metallic enclosure. This metal enclosure may be freestanding, attached, or integrated to the building elements. It can even be in the form of copper based paint.

The Department of Defense (DOD) often will require a passageway directly outside the metal shield enclosure to provide a means for inspecting the integrity of the enclosure.

The shielding envelope must be continuous, free of openings, which could allow leaks. This can pose some unique problems in the treatment of windows, doors, air duct openings, and plumbing and electrical penetrations. There are techniques and devices available that address all of these areas. There are windows that utilize metal mesh and conductive metallic coatings, doors with special gaskets, grills that permit the free flow of air but impede electromagnetic waves, and electrical filter fittings for plumbing, etc.

It is important to incorporate a positive method in joining the metal shielding components to provide an uninterrupted barrier. Seams must be tight-fitting, free of nonmetallic paint or other coating such as dirt, grease, rust, or other insulating material. Various techniques can be employed, including welding, soldering, mechanical fasteners with pressure plates, and conductive tape.

When an electromagnetic wave encounters the enclosure's electrically conductive material, the portion of the wave transmitted beyond the shielding barrier is reduced in magnitude by both reflection and absorption. The reflection loss occurs at the two interfaces between the transmitting medium (air) and the shielding metallic material. Absorption takes place as the wave passes through the conductive material. Absorption losses occur from dissipative heat loss caused by currents induced in the metal.

An electromagnetic shield can be defined as any barrier which reduces the level of the electromagnetic field. How well the shield attenuates the field is referred to as its shielding effectiveness (SE). The standard unit for expressing the ratio of two amounts of electric or acoustic signal power is the decibel (dB). It is the unit for measuring shielding effectiveness. The overall shielding effectiveness of any barrier will depend on the particular type of electromagnetic field being emitted and the particular material selected as the shield.

The purity of electric current passing through wires and

cables or transformers is dependent on the quality of the shielding provided on those wires or cables. Conditioned power and data transmission lines often require special shielding to preserve the quality of the signal being transmitted.

Radionuclides

There is a growing need to provide shielding around X-ray equipment or irradiated materials to contain the strong emanating energy of X-rays and radio isotopes. Nuclear radiation and X-ray exposure can cause serious damage to live tissue and must be contained for safety reasons. Plexiglas is an efficient barrier to low level isotope radiation, as is lead (Figure 13.1). So-called "hot" labs where P-32, I-125, and other radiated materials are manipulated in living tissue can have the room partitions faced with .25" thick Plexiglas to contain the radiation. Lead can be used, but it is much more costly. A much denser material such as lead is generally used for shielding X-rays, cesium, and cobalt radiation, or heavily reinforced concrete enclosures with a baffle screen maze entrance of thick concrete are provided for shielding cobalt irradiators.

Radiation-producing equipment must conform to the Radiation Control for Health and Safety Act, and any implementing regulations issued by the Public Health Service of the Department of Health and Human Services, Nuclear Regulatory Commission Standards for Protection Against Radiation (10 CFR 20), and guidelines issued by the Environmental Protection Agency under "Guidance for Occupational Radiation Exposure."

Magnetic

The unit of measurement for a magnetic field is a gauss. This is the unit of magnetic force alone, without the electric energy component. On planet Earth we are always subject to at least .5 gauss due to the earth's natural magnetism.

Figure 13.1.   Plexiglas radioactive waste containers. The Plexiglas provides a shield to contain the low level P-32 radionuclides.

Modern technology has enabled man to create very strong
magnetic fields that can even change the random alignment
of living cells in a body to a unidirectional alignment. These
cells can be observed with the use of special imaging computer equipment to detect anomalies.

Strong magnetic field research is not limited to medicine.
There is a growing use of magnets in many areas of modern
research. People with iron alloy metals embedded in their

body, such as a pacemaker, shrapnel, orthopedic pin, or fill-ing can be seriously injured if subjected to a high energy magnetic field.

Shielding magnetic resonance imaging machines requires two layers of 1 inch thick special alloy steel welded plate en-closures to contain the magnetic field emitted from that type of equipment; and even then, color video display monitors can malfunction as far as 20 feet beyond the shielded enclo-sure without additional supplemental shielding.

There are specialists in shielding design. The architect should seek the advice of a shielding consultant where mat-ters of personnel safety, proper equipment operation, or in-formation protection issues are subject to compromise with-out appropriate shielding.

14

Plumbing Systems

Plumbing systems are essential infrastucture for "wet" labs for glassware and equipment washing, solvent/dilution service, and pure water distillation ranging from sterilization and bio-kill applications to water for injection (WFI). Modern laboratories have several different water systems including: domestic system, "protected" water system dedicated to research activities, pure water for product processing, and water for injection where the product may be injected into humans. Sterilization uses water as a medium for nearly every process; heated in glass lined or stainless steel heat exchangers, the material to be sterilized is sustained at a high temperature for sufficient duration to kill any microorganisms.

As with heating, ventilating and air conditioning systems configurations, the plumbing systems layouts must consider flexibility and capacity for future changes. Ease in reconfiguring laboratories and limiting disruption of adjacent space are primary goals for appropriate lab design. This is often difficult because lab drains are generally dependent on gravity flow. Emergency shower and eyewash stations are a code requirement for chemical laboratories. The possibility of adding dangerous microorganisms to the waste stream is also a concern to the plumbing system design engineer.

Water Supply Systems

Many plumbing codes require a complete separation between the domestic water system serving the building and the water service for the laboratories. This so-called non-potable "protected" water system used in the pharmaceutical and biological research environment is generally supplemented by some form of pure water system. This can be distilled water which is usually deionized as well. Some levels of research will require reverse osmosis deionized pure water (RODI) systems, salt tanks, settling chambers, and sometimes ultraviolet purifiers. Pure water systems must be continuously circulated so that the water does not become static in the piping system and subject to contamination. Consequently, dead-end leg lengths are kept to a minimum and the initial pure water coming from the outlet is drained away and wasted for several seconds prior to actual use. Super levels of purity beyond RODI are generally not accomplished using a central system. Super purity is best accomplished with a local device attached to the central pure water system at a specific lab bench location (Figure 14.1). The most common grades of pure water for laboratory applications are:

- Reagent grade
- United States Pharmacopoeia (USP) purified
- Water for injection (WFI)
- Electronic grade (highest standard)

According to the quality of the available domestic water source the systems and components required to achieve the desired levels of purity will vary; i.e., softening may or may not be required in addition to other forms of 'polishing.'

Piping

The purer the water, the more aggressive it is; i.e., the greater its ability to dissolve piping components. Conversely, piping vessel material can have an effect on water quality. It is, therefore, important to select the appropriate pipe material for the fluid to be used. Among the piping materials used in labs are:

Figure 14.1. Pure Water Final Polishing System, mounted above lab bench sink.

- block tin
- tin lined copper
- stainless steel - for process supply
- polyvinylchloride (PVC)
- polypropylene (PP)
- glass - for radioactive wastes
- aluminum
- polyvinylidene fluoride (PDVF)- for ultra-pure supply

Gas, Vacuum Systems

Natural gas, compressed air, and vacuum outlets are generally provided for lab bench work and they are usually distributed from a central system. Other gases are frequently required for specific equipment or processes, but they are normally supplied locally from replaceable cylinders (Figure 14.2). Nitrogen and carbon dioxide fall into this category. The handling of gases may require ultra clean distribution systems with flow regulators, purifiers, analytic devices, and/or wet chemical scrubbers. To ensure safe handling of gas, custom designed gas cabinets are often required for leak purging, fireproofing, and use monitoring.

Liquid Waste

Selecting materials for research facilities means more than simply meeting current code requirements. It requires an analysis of the types of chemicals to be discharged into the system and their effect on different piping materials. A drainline and neutralization sump system must protect people, buildings, and equipment from dangerous exposures, and assure that discharges do not exceed local municipal sewer requirements. These safety considerations make it imperative that drainlines and ceramic or concrete neutralization sumps maintain physical integrity over the years to ensure leak-free service.

In addition, with the increasing use of radioactive imaging, biological research labs deal with low-level radioactive

Figure 14.2. Carbon dioxide tanks—4 active and 4 standby tanks serving remote incubators in tissue culture rooms. When active tanks become empty, automatic change-over device activates the standby tanks and sounds an alarm to notify lab personnel to replace the empty tanks.

wastes (radwaste). The type of drainline required for rad-waste depends on the type and level of radioactivity.

Plastic materials are unacceptable for radwaste applications because even weak radiation may, over time, destroy the cross-linking in the polymers, weakening the pipe and allowing unacceptable levels of radwaste to escape, usually in concealed spaces above the ceiling of occupied space. Glass drainlines are unaffected by radiation. In fact, glass is considered ideal as a long-term encapsulation material for high level radwaste. Taking into consideration the ease of installation and future modification, ease of maintenance, lower life cycle costs and safety, the use of glass drainlines certainly must be considered. Unlike plastic or metal pipe, glass requires neither solvents nor heat welding for installation. Glass pipes are joined with stainless steel mechanical couplings, connecting either bead to bead ends of pipe, or bead to plain ends. Only a socket wrench is needed to make the connection. Glass pipe is unsurpassed in corrosion resistance against the widest range of chemicals, acids, low level radiation, and contaminants found in laboratory drain and vent systems.

Glass pipe is also used for high purity water systems. It is chemically inert, allows less than 0.08 parts per million pick-up of contaminants, and incorporates jointing components and methods that are resistant to chemical contamination. Glass pipe is resistant to temperature extremes with a low thermal coefficient of expansion which allows safe handling of liquids up to 212°F continuous and 250°F intermittent. Unlike plastic piping, often used in pure water systems, glass will not emit toxic fumes or provide fuel to a fire.

Laboratory waste cannot empty directly into a municipal sewer sytem since it can contain acids, harmful bacteria, microbes, and pathogens. The acids are neutralized in sumps filled with either marble or limestone chips. The natural chemical reaction between the calcium of the stone effectively raises pH and neutralizes acid wastes in the tank. Where it is impractical or inadvisable to install a neutralizing sump, a pH adjustment tank should be provided. The pH adjustment tank is constructed of acid/alkaline resisting material with an agitator which mixes the wastewater and a sensor that detects the pH of the tank content within a range of 2 (acid) to 12 (alkaline). The sensor connects to automatic electronic controls that operate pumps for acid or alkaline neutralization agents to bring the tank to the 6 to 9 pH range. The local wastewater regulatory agency may require an outflow recording device equipped with an audio-visual alarm. Municipal regulations can require an automatic shutdown in the event of improper waste neutralization operation and compliance documentation. Effluent streams should not be heated. Most municipal sewer systems will not permit effluent to be above 120°F (48.9°C).

Acid neutralization sump tanks can be monolithic ceramic stoneware, steel shell, or reinforced concrete. A reinforced resin coating can be applied to prevent sump corrosion when handling concentrations of hydrofluoric or strong caustic solutions. Often a second sump tank is necessary to further neutralize the acid content. This second tank generally contains soda ash, a magnesium carbonate that further raises the pH level. Biomedical labs need secondary stage treatment tanks of chlorine to act as "kill" tanks to treat-sterilize any bacteria and pathogens that may be present in the liquid waste.

Most codes prohibit viable organisms containing recombinant DNA (deoxyribonucleic acid), as defined in the National Institutes of Health (NIH) guidelines, to be introduced into the lab drainage system or sewer without first being sterilized, treated, or inactivated. Code or no code, it would be imprudent to ignore these organisms. The NIH Recombinant DNA Guidelines and the Laboratory Safety Monograph should be followed for biomedical and production laboratories.

Waste containing recombinant DNA organisms should be sterilized or treated at the point of origin or, where there is more than one point of origin, these wastes may be collected in a central holding tank for sterilization and treatment. The tank should contain a sampling device and a high water alarm. Before the contents of the tank can be allowed to empty into the sewer system, the sample must be determined to contain no living organism. Retreatment or sterilization to the tank contents must be performed if the sample shows live organisms.

Passivation

Process piping used in the biopharmaceutical industry must not react detrimentally to the fluids contained in the pipes. Historically, type 316L stainless steel must undergo a passivation procedure using nitric acid as outlined in ASTM A380. This can be a hazardous and expensive process. Nitric acid is a fuming, suffocating, and corrosive liquid; its fumes are very toxic and the liquid causes severe tissue burns. The exposure limit for nitric acid fumes is very low—a peak short-term (15 minutes) exposure limit is 4 parts per million. All noncritical personnel must be evacuated from the premises prior to the procedure which calls for 35% nitric acid, heated to 140°F to be circulated in the piping system for 30 minutes. Any leak could cause a disaster.

The passivation procedure coats the inside of the pipe with a very thin (25 angstoms), relatively unreactive chromium oxide/hydroxide-enriched passive surface film which is very resistant to corrosion, even from very aggresive pure water. The Food and Drug Administration requires water for injection (WFI) piping to be passified and validated prior to use.

Recently, substitute chelants for nitric acid have become available in the form of nontoxic, biodegradable chemicals which are much safer and more effective.

Costs

Plumbing systems in a modern "wet" lab contribute about 10% of the new lab building total construction cost, when excluding site improvement costs. For lab renovation work the percent of construction costs associated with the plumbing systems is usually somewhat higher than for new work.

15

HVAC Systems

Designing laboratories is a complex undertaking—labs are not just any other occupational space. The circulation/ventilation air for heating and cooling, fume and odor exhaust, particulate control, pressurization, and make-up air supply are the components of the lab air systems. Research laboratories are expensive to construct and operate. The demands for exacting environmental conditions required by the research community translate into sophisticated HVAC systems, high energy consumption and high operating costs. It is normal for labs to consume between 300,000 and 400,000 Btus per square foot per year. This is easily 6 to 8 times the number of Btus consumed in the modern office building. Properly designed environmental systems must help prevent contamination and distortion to experimental data, maintain appropriate temperature, humidity, and air flow conditions, and facilitate a safe workplace for the building occupants and the surrounding community.

Certain types of pharmaceutical research require sterile environments with Class 100 and Class 1,000 clean rooms in order to meet Food and Drug Administration (FDA) and/or Good Manufacturing Practices (GMP) standards. Nevertheless, every effort must be made to consider practical energy saving devices such as air to air plate heat exchangers and/or automatic DDC systems controls to tune the HVAC systems use to the ever-changing occupancy needs. The American Society for Heating, Refrigeration and Air Conditioning Engineers (ASHRAE), in Chapter 14 of its Applications Handbook, clearly spells out the design information needed for research labs. Importantly, it emphasizes that each research lab is unique in its design and "must be evaluated using current standards and practices rather than duplicating the designs of outmoded existing facilities."

Interestingly, the ASHRAE standard points out that labs for low or moderately hazardous work have greater exhaust air needs than ones performing more hazardous work. The reason for this seeming paradox is that labs performing more hazardous work usually do so in devices that provide greater protection, such as sealed glove boxes, which require far less exhaust air than do lab fume hoods and biological safety cabinets.

Biomedical laboratories and their "wet" support areas should have 100% fresh air and 100% exhaust. Four to six outside air changes per hour are recommended for a "safe" BL-2 lab and lab support space. In areas of labs that routinely use carcinogens, 10 outside air changes per hour are called for. Recirculation of air locally within a lab or lab support space (fan coil or heat pump) is acceptable, but not from one room to another. Ceiling return air plenums are not a good idea. Lab return air should be hard-ducted. There should be no opportunity for cross contamination of return air via common returns for air handlers serving labs and lab support spaces. Of course, offices and conference rooms may have a recirculating air system.

A conscious effort is necessary to minimize drafts in certain critical areas of a laboratory. Microtome work, the thin slicing of tissue, cannot be performed in areas of even moderate air movement. The chemical balance area is another vulnerable work space that should not have drafts for obvious reasons. Bench work often involves precise measurements and good hand/eye coordination. Excessive air movement can inhibit these activities.

Pressurization

Maintenance of positive and negative air pressure relationships among various types of laboratory spaces is essential for the prevention of biological, chemical, and/or particulate contamination. This requires careful planning, construction, and operational practices to direct laboratory air flows to predictable paths. Such pressure controls require substantial HVAC capacity to respond to sudden changing conditions when doors are open during times of materials delivery and personnel movement. Approximately 250 cfm of air should flow away from any lab entrance.

Filters and Exhausts

Designs for laboratory exhaust air requirements must consider the type, number, size, and frequency of fume hoods and other particulate containment devices. Specific design criteria include factors associated with the:

- cost to build and operate
- safe management of expected hazards
- research productivity
- flexibility in adapting to research program changes, and
- matching safety features to assessed research risks.

There is always the potential for airborne contaminants wherever biological materials are present. The design of the building exhaust systems and filters for ducted air systems becomes crucial to the safety of the occupants and neighboring community, and to the reliability and efficiency of the scientific research. Certain types of research may require incineration of the exhaust particulates to ensure that no living organisms leave the building. Exhaust system design will vary according to the scientific studies being conducted. Certain BL-2 level work requires only minimum air exit velocity with no treatment. Air sampling devices are required with the potential of certain irradiated isotopes in the effluent. Single chamber incineration may be required for some airborne organisms and dual chamber exhausts with incineration combined with caustic scrubbers are often required for pathogenic material. In any event, the exhaust air should exit the building at the top and be remote from the fresh air intake nostrils.

Filters are usually placed across the intake and exhaust air paths. They are often introduced within the labs to enhance the requirement for sterile environments. Activated charcoal filters are used for gaseous contaminants. High efficiency particulate air (HEPA) filters are often used in biological labs. Sometimes, HEPA filters are light scanned monitored to detect possible flaws in the filters where extremely sterile conditions are desirable.

Planning a facility using bypass hoods, the designer need plan for only one supply ductwork and air-handling system; namely, that needed to serve the room with the fume hood. However, it must be sized to make up 100% of the hood exhaust.

In a facility using make-up air type fume hoods, two systems must be provided: one serving the room with heating and cooling capabilities and one serving air directly to the hoods needing only heating capabilities. Potentially, with the make-up air hood, requirements for conditioned room air can be reduced to 30%, thereby diminishing capacity needs drastically. Obviously, these are important planning issues affecting space requirements and central utilities. Decisions on these issues must be made early in the planning process.

Waste storage rooms should have independent exhaust vents directly to the exterior. In a lab, most waste should be short-term; they are: animal waste, biohazardous (red bagged), and chemicals.

Temperature and Humidity Control

Air supplied to a laboratory is drawn from outdoors, prefiltered, filtered, heated or cooled according to demand, humidified or dehumidified according to needs, and distributed via fans and ducts to maintain the proper pressure relationships among the various lab spaces. This supply air replenishes the air that becomes contaminated within the labs and is automatically exhausted through chemical fume hoods, wash area hoods, and certain types of biological safety cabinets. Laboratory air always moves from clean to dirty, positive pressure to lower pressure, and from supply or "make-up" source to exhaust. It is important to place the exhaust air stream away from the supply intake louvers to prevent re-entrainment of contaminants.

Direct digital controls (DDC) provide the capability to operate each separate zone or space at individual temperature and humidity set points. Each zone can be monitored and reset manually or automatically, utilizing software that will respond to historic preaction, optimization of conditions, alarm, and abnormal notifications.

Basic design temperature for most laboratories is 70° to 76°F. In seasonal climates, the relative humidity can fall to 25% in the winter season, which is not ideal for human comfort and for some types of laboratories where static electricity can cause problems with the research tasks. Artificial humidification should be considered for ideal laboratory winter conditions. The natural dehumidification which accompanies the air conditioning system during the summer season ordinarily maintains a comfortable RH level. Humidification is a must for animal facilities and environmental warm rooms.

Estimating the heat loads for a laboratory is crucial to the design of the air conditioning capacity. Besides the normal heat gain calculations for the building skin, the interior heat gain from lighting, people, and especially lab equipment must be considered. Lighting loads for offices and conference rooms normally run in the 1-2 watts/SF range. Laboratories tend to have greater lighting loads, averaging in the 2-3 watts/SF range. Laboratory power requirements can vary significantly. Biomedical labs will use 8-10 watts/SF, except for certain equipment rooms which may be as high as 40-50 watts/SF. On the other hand, electro-optical labs can average 20-25 watts/SF, with some concentrations at much higher levels. In order to select the most appropriate central air conditioning equipment, the engineer must carefully analyze the anticipated total load and the expected diversity.

Representative heat outputs for various "wet" lab generic equipment are as shown in Table 15.1.

Equipment housed in communications closets can gener-

Table 15.1. Representative Heat Output of Lab Equipment

Autoclave/Sterilizer:	14,700 Btu
Drying/Sterilizing Oven:	11,400 Btu
Glassware Washer:	12,000 Btu
Glassware Dryer:	10,000 Btu
CO_2 Incubator:	700 Btu
Upright Incubator:	4,100 Btu, 60% diversity
Biosafety Cabinet:	1,500 Btu
Upright Freezer:	1,500 Btu
Upright Refrigerator:	1,500 Btu
Ultracold Freezer:	3,000 Btu
Scintillation Counter:	4,000 Btu, 60% diversity
Ultracentrifuge:	7,000 Btu, 80% diversity
Low Speed Centrifuge:	10,000 Btu, 80 % diversity
High Speed Centrifuge:	14,000 Btu, 80% diversity

ate significant heat. The temperature control and ventilation for these spaces are critical. Computer communication and local area networks are important infrastructure for successful and productive research activities. Some of this communications equipment is heat sensitive and can fail under the wrong conditions.

Costs

Excluding site development costs, HVAC systems and their controls account for 25% to 35% of the construction cost of a new state of the art "wet" laboratory. This represents more than the cost of the building excavation, foundation system and structural frame. For renovation work, this percentage can increase to as much as 50% of the total construction cost. Because of the high initial and operating costs related to the large and complex air movement systems related to laboratories, more and more designers are recommending that a diversity factor be considered when sizing the central systems capacities. Studies and practical experience have shown that for large labs with many chemical fume hoods, at least 20% to 30% of all exhaust ports will be closed or only partially used at any one time. This allows the central equipment to be sized for 70% to 80% of the peak ventilation capacity, resulting in a substantial real cost savings. Of course, space is usually provided for future equipment in the main mechanical spaces to meet any eventuality. However, nothing is without risk. The building operating staff must be instructed about the maximum capabilities of the diversity-driven design so that careful energy saving procedures occur on the summer peak demand days.

16

Electrical Systems

Research laboratories must have electrical service and distribution systems designed to provide reliable, flexible, uniform, and sufficient power with good access to equipment and wiring to accommodate future changes. To achieve these goals, system modularity must be given full consideration in developing design concepts. The modern research facility is not only dependent upon electrical power, but also on efficient multichannel FFDI fiberoptic cabling for computer data transmission, or twisted-pair wiring for telephone, coaxial cable for TV, and other kinds of low voltage wiring systems for security and audio-visual equipment.

Power Reliability

Living cells involved with long-term experiments are stored at extremely low temperatures. Consequently, any power disruption to the low temperature freezers can destroy expensive, long-term scientific research. Certain other critical devices such as fume hoods, cold rooms, fans serving sterile environments and animal facilities must have adequate standby generator power to protect the investment in experimentation and building occupants.

It is very desirable for the research laboratory to be served by multiple feeders operated in parallel. Ideally, incoming power distribution should consist of more than one utility service feeder or substation with an interconnecting tie for a higher degree of reliability and flexibility. Then, systems can be sized such that, with the loss of one service, the remaining systems can accommodate a significant share of the total demand load. This setup will allow continued research operations during times of utility company transformer failure or storm damage, without overextending the emergency generators. Emergency power should include both clean and commercial grade power systems.

Systems Flexibility

Given the nature of research with its constantly changing equipment demands, flexibility should always be considered. The power distribution systems must be designed for both short- and long-term needs. The quantity and quality of power, frequency, and voltage levels are all important. Among the items to consider are:

- known powerloads and likely future demands
- load diversity
- spare distribution equipment
- bypass/isolation switching
- space for future growth
- modular layout
- organized wire management.

The designer should allow individual rooms sufficient service capacities to handle localized loads, but not oversize the central system unnecessarily. Typically, numerous electrical outlets are placed throughout a lab, allowing the scientists to place apparatus at convenient locations. Generally, only a small number of outlets are in use at any one time; however, the arrangement allows possible localized high intensity power demands. Establishing the diversity factors for branches, submains, mains, and central equipment becomes a crucial decision to be shared among the designers, researchers, and facilities managers. Obviously, initial project cost can be greatly affected by the levels of diversity chosen.

It is not unusual for some laboratories to operate with 75% of the frequently spaced outlets to be using power. However, it would be an unusual building where all of the labs would be drawing from 75% of all outlets at the time of peak demand. A more expected overall laboratory building diversity factor would probably approach the 25% to 30% range. In other words, the low intensity labs in the building average with the high power users to lower the overall power diversity. In a modular lab building design, the outlets per lab module can be standardized for future flexibility. The actual power demands for each individual lab will probably vary over time, yet the actual total building power consumption will probably grow at a relatively slow rate.

When specific lab equipment information is not available, the designer should consider providing each generic lab module with nearly all of the services required for the most power intensive lab to maintain future flexibility. Services not required initially should at least consist of empty conduits terminated at chases near the labs, and with panel board space reserved for future branch circuits. It is much simpler to deal with future circuiting needs in an occupied and functioning laboratory in this manner than to install additional feeder conduits at a later date. This philosophy for flexibility applies to the building core areas and lab equipment support spaces as well.

Equipment or instrument rooms, which are often self-contained to isolate the noise and heat gain, are a special challenge to the electrical design engineer. The equipment in these spaces will likely change many times over the life of the building. Therefore, various outlet types with varying voltages and configurations must be considered throughout the area. Data connection points to the local area network for instrumentation monitoring in the instrument rooms are generally required.

The main conduit power feeder runs and low voltage cable trays should not run above the lab ceilings. They should be grouped above the corridors or office areas to avoid future lab disruption during maintenance and minor renovation work.

Power Quality

Clean power distribution systems are often necessary in a modern research facility because of the reliance on computer instrumentation. Power serving sophisticated laboratory equipment must be free from transients, harmonic distortion, and ground currents. The development of new technologies in electronic equipment over the past decade has raised concerns about power inconsistencies. There was little concern about power variances until the past 10 to 15 years. Power systems have always been designed to run linear loads such as heavy motors and lighting. The introduction of microprocessors into the workplace has changed that. Now, there is a demand for better power quality.

Power problems include the following:

- Improper or inadequate grounding results in EMI/RFI noise.
- Transients, spikes are pulses that can be only a few nanoseconds duration, but can be 6KV, 3KA.
- Voltage irregularities such as brownouts or overvoltages for longer than half a cycle.
- Harmonics are distorted sine waves caused by current on the neutral wire generated by switch mode power supplies.

The reality is that most utility companies generate good quality power, but weather, other power users, internal demands—even traffic accidents and rodents—can cause spikes, dips, and noise that can destroy or severely disable sensitive electronic equipment. It is the fluctuations, more than the massive power failures, that make the need for line conditioning critical. The following power conditioning devices may be required to provide adequate power uniformity:

- surge suppressors
- isolation transformers
- electrostatic shielding
- voltage regulators

- ferreosonant conditioners
- UPS system with battery backup
- lightning protection

It is estimated that about 20% of power quality problems can be classified as true power line irregularities; 20%—a deceptively low probability of power disturbances damaging or destroying electronic data and equipment, but certainly a risk that should not be taken, in any probability. The remaining 80% comes from a combination of faulty wiring and improper equipment layout. Voltage fluctuations are becoming an ever-increasing problem with varying internal and external demands on supplied power.

Power Conditioning

A good line conditioner combines voltage regulation, isolation to filter noise, and surge suppression to eliminate

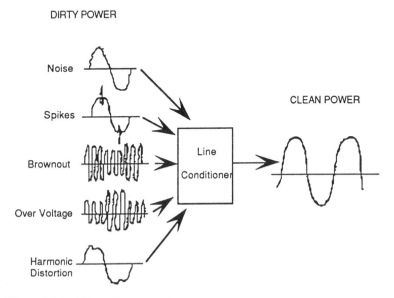

Figure 16.1. Alternating current.

damaging spikes (Figure 16.1). A line conditioner with these elements will provide clean, consistent power under most conditions, short of a total power failure. Line conditioners are available using several technologies. These include isolation transformers, voltage line regulators, tap switches, and the new ferreosonant technology which eliminates switching delays. The question becomes how best to protect electronic equipment from the inherent fluctuations in, and the periodic loss of electric power.

Ground Fault Circuit Interrupters

Unlike fuses and circuit breakers that are designed to protect circuit overloads, ground fault circuit interrupters (GFI) are designed to trip when an imbalance of only 5 to 8 milliamperes occurs between wires serving and returning from electrical devices. Therefore, the circuit interruption will protect personnel from lethal doses of electric current when the person has accidentally grounded the circuit. GFI devices should be installed wherever portable or nonstationary lab equipment will be connected. At lab benches there should be a maximum of 3 duplex outlets for each GFI. Because of the increased potential for accidental grounding, GFI devices should always serve outlets near a sink.

Lightning Protection

Thunderstorms can cause drastic and damaging spikes to power distribution systems. Consequently, the building should have an Underwriters' label rated lightning protection system. The purpose of such a system is to prevent lightning strikes to the building by having frequently spaced metal points connected directly to grounded cable which will allow the electrons to escape and neutralize the electromagnetic energy or static differential between the building and the cloud mass.

Uninterruptible Power Source (UPS)

Uninterruptible power systems (UPS) generally only serve computer rooms and critical laboratory instruments, controls, and monitor. There are three basic types of UPS systems currently available—off-line (SPS or standby), on-line (UPS), and hybrid (SPS with line conditioning).

In an off-line or standby system (SPS), raw alternating current power feeds directly into the equipment. The invertor turns on only after a power outage occurs and power is drawn directly from the battery. Generally, it takes more than 4 milliseconds for the transfer switch to activate the invertor. This transfer time is critical because not all loads can tolerate that transfer time. An SPS generally does not provide protection from sags and surges, or over and under voltage conditions, and may or may not provide surge protection. While the cost of an SPS is lower than other systems, the cost should be weighed against the level of protection the system provides.

The hybrid design battery backup system is basically an SPS with addition of a line conditioner on the output. This system provides some spike protection that a basic SPS does not provide. Like the SPS, the invertor is off during normal operation and only activates after a power outage. Unlike the SPS, during the transfer, the "hold-over" of the line conditioner will continue to provide power to the load. However, the output will experience a voltage dip and phase shifts are possible which some loads, disk drives for example, do not tolerate well.

On-line (UPS) provides the highest level of protection from all power disturbances. In an on-line system the invertor is always on, providing clean regulated power to equipment. If the alternating current power fails, the invertor simply draws from the battery with no transfer switch and absolutely no interruption in power to the load. While initially more expensive, an on-line UPS is the best insurance available against a power failure.

Usually, a UPS is coupled with an emergency generating system so that the batteries only have to carry the total load during the time it takes to start the generators. This time is generally up to 3 minutes. The size, number, and condition of batteries on a UPS determines the length of time and size of load they can carry during a total power failure. For maximum protection, an on-line UPS with an emergency generating system is recommended. There is no switching time, or power gaps, between primary power source and the batteries. The UPS equipment also provides precision frequency control, which removes the threat of phase shifts by being on-line all of the time. Phase shifts occur when transferring from one power supply to another, i.e., utility power to battery in SPS or hybrid systems where the line conditioner is, cannot prevent phase shifts. UPS battery backup facilities are quite expensive to build and maintain when the equipment loads are significant.

Sufficiency

Research laboratories can have vastly different power requirements. Specific levels can range from 10 watts/SF to 100 watts/SF, depending on the equipment involved. Because lab layouts are constantly being reconfigured throughout the life of the building, designers must try to predict and understand the long-term projected needs of the technology.

Each station or five foot length of bench in a modern lab needs the standard of one double duplex, 120 volt alternating current (vac) 20 amp commercial power receptacle. Each lab bench consisting of several lab stations should have a telephone outlet; a double duplex, 120-vac, 20 amp emergency power receptacle; one duplex 120-vac, 20 amp clean power receptacle; and one duplex 120-vac UPS receptacle.

The electrical systems for a modern research facility are varied and complex. Questions need to be answered about equipment and operations that will affect various outlets for power, communications, and data. Wherever possible, the concept of modularity in systems design should be considered for facilitating future changes. Labs with areas of research using hazardous substances need to have adequate emergency power to protect the scientist in the event of a power outage. If high voltages (5kV and higher) are to be distributed, electronic interferences will require considerable care to isolate and shield the cables. Appropriate locations for electric panels need to be established for ease of accessibility.

A modular-designed research facility with standard repetitive rooms that contain both benches and fume hoods should also include additional auxiliary power receptacles such as 120-vac, 20 amp commercial power outlets; a 208-vac, 20 amp emergency power receptacle for lab freezer, a duplex 120-vac, 20 amp outlet at each fume hood; at least one lab automation outlet (RS232 or equvalent) for connection to a local area network; a monitoring alarm system connection; and one 120-vac, 20 amp commercial power receptacle below each lab sink for maintenance purposes. In addition, a panic button should be provided to "kill" the commercial power in the area for safety purposes.

Additional custom electric services are generally provided for specific lab equipment. It is important to label and color code the various types of receptacles so the researchers can readily identify the appropriate power outlet for their needs.

There should be provisions for 480 vac commercial and 480 vac emergency power systems serving distribution panels on each floor in centrally located electrical equipment rooms. Each of these panels will serve lab equipment requiring 480 vac service and low voltage transformers for 120/208 vac panel boards. The location of these equipment rooms and distribution panel boards is important to the overall floor layout, and floor space needs to be dedicated early in the design process. It makes sense to locate these spaces near the vertical chases which are often shared with other services.

Having regular spacing and consistent locations for the lab circuit panels makes sense for the users. Each electrical closet should be similar in size and configuration, even though initially the number of panel boards installed may vary due to different usage densities. If a 42-pole panel board is selected, the first 30 poles should be dedicated to serve typical receptacles plus spares. The last 12 poles can be devoted to serving the custom or nontypical requirements of the area served by the panel, and to provide for future special needs.

Backup or emergency electric power is required for fume hood exhaust systems in labs of BL-2, BL-3, and BL-4. Animal facilities should also have an alternate power source for their ventilation systems. Environmental conditions, such as temperature, pressure, ventilation, etc., in areas serving expensive, hazardous, or volatile materials are critical. Alternate power should also serve low temperature freezers, cold rooms, selected laboratory instrumentation, computing equipment, and of course, exit lighting and other life safety systems.

Lighting

Natural lighting is very desirable for generic bench space, provided there is no glare. Artificial lighting controls are important in laboratories. Many special work environments in a laboratory do not want natural light because of the level of lighting controls needed. Some spaces, like darkrooms for film developing, should not have fluorescent lamps due to the "ghost" radiation that is emitted after the current has been turned off. Some labs require light wave lengths to be filtered in order to eliminate certain visible light wave frequencies. Semiconductor research labs often have yellow lighted rooms without any blue, green, or violet components of light. Dimming controls are often desirable as in microscope rooms. In addition, artificial lighting systems should include normal, night lighting, emergency, and egress lighting. Multilevel artificial lighting controls are desirable to give the individual scientist control of the quality and intensity in the labs and office areas. This is especially crucial where video displays are used.

Security

Electronic powered security systems are necessary for research laboratories. Closed circuit television (CCTV) surveillance, card access, door monitoring, and remote controls to access restricted areas are all commonly used systems. Alarm systems that monitor research instruments to report trouble or upset conditions have become commonplace in research facilites. Automatic fire detection systems are now required by all municipalities.

Local Area Networking

Communication systems include voice (telephone and intercom), television, and data. These networks are essential for communicating and recording information quickly and accurately. These systems outlets need to be ubiquitous in a modern laboratory where nearly every researcher needs a handy computer terminal.

Costs

To illustrate the dependence of modern scientific research on electricity, it is not unusual for a major facility to pay a million dollar electric utility bill each month. The electrical systems construction cost for a new laboratory are normally in the 10% to 15% range of the total building construction cost, excluding site work. The peak power demand for a single 15-minute period each month is a big factor in determining the amount of most electric utility bills. Energy management should be a major consideration of laboratory design and operating procedures. The air conditioning demands resulting from the enormous volume of air exchange requirements of a lab are a major contributor to the high use of electric energy.

17

Fire Protection Systems

Laboratories should always have an automatic fire suppression system designed in accordance with NFPA standards and connected to a Class A supervised central alarm. Wherever unusual hazards exist, special systems are appropriate. For example, elemental sodium will react violently with water. The local fire official and a competent safety engineer should be consulted as to the best system to employ with the specific special hazards.

Handheld portable fire extinguishers are also necessary for a safe laboratory following the NFPA guidelines. Carbon dioxide is usually the preferred media, with 5 pounds the minimum practical size. They should be placed both remote from the lab door and near the lab door for easy access. If possible, first aid fire hose cabinets should be avoided for a laboratory, for they can cause havoc if operated by a novice in areas of unusual chemicals. On the other hand, smoke detectors tied to a central fire alarm system can protect lives and property in an environment where experiments often go unattended around the clock.

The cost of automatic fire suppression systems are a bargain, considering the safety benefits and reduction in costs for insurance premiums. These costs usually run less than 2% of the total construction cost for a new, state-of-the-art research laboratory.

18

Security

Issues

In a research laboratory, the need to protect property goes hand in hand with the need to protect information. Labs contain highly sensitive and expensive equipment, and ongoing experiments that could be easily destroyed by inadvertent access. Secret data, if disclosed to the wrong party, could compromise a company's competitive advantage or the nation's defense. The investment in a variety of research animals is at risk should some animal rights zealots try to access the facility to "free" them from captivity.

Strategies

There are four aspects to facility security: manpower, physical barriers, technology, and procedures. A good security plan deters intruders in the first place, detects their presence should they get into an area where they don't belong, and, finally, delays achievement of their devious deeds. This is accomplished by a careful balance of thoughtfully designed buildings, a well trained staff, and the right technologies for deter/detect/delay responses. Optimum security for laboratory employees and research programs is achieved through an integrated program. An increasingly important part of the integration process is coordination with the user's concerns, such as the security objectives, and their functional imperatives. The design team must understand the need to identify and validate visitors, to deter, detect and respond to unauthorized entry before attempting to design physical entry sequences.

For good reason, you will seldom find a building directory displayed in the lobby of a research facility that tells the visitor where to find the cobalt irradiator or announces that the facility contains a vivarium with holding rooms for primates or rabbits. The visitor to a reseach facility is usually required to verify their need to be on the premises, to have that purpose validated by a responsible party within the facility, to log in and out of the building, and to be accompanied to their destination within the building. Often, this is more for their own safety than to protect secrets. Labs can be dangerous places for the uninitiated. In certain defense related research facilities a flashing yellow strobe light announces to the research staff that there is a visitor on the premises and any sensitive information or activities should be concealed, including video display screens.

Building perimeter design and internal controls should reflect real security needs. Legitimate access should be simple to operate and cause a minimum impact on day-to-day facility operations. In addition, exposed hardware should be selected to blend with the building aesthetics rather than resemble standard military issue. Modern technology and miniaturization have made great strides in the appearance of items such as cipher locks, card readers, and cameras.

Electrified locks are now commonplace. Selecting the proper type for the particular condition requires some expertise. Some of the don'ts include:

- Don't use electric strikes on reentry fire stair doors. Building codes require the stair door to remain latched (but not locked) in the event of a fire.
- Don't use electric deadbolts on doors in the path of egress. Pressure on the door will not allow the bolt to retract.
- Don't use electromagnetic shear locks on wide or heavy doors, or on any door where deflection or settling will likely occur. The strike plate will not close with the magnet, resulting in an unsecured door or loud sounding lock, and possible burned-out power supply.

It is best to seek the advice of a qualified hardware or security consultant to determine the most appropriate applications for the situations to be encountered.

SCIF

Defense related research programs often include special facility design requirements and operating procedures dictated by security concerns. Secure, Compartmented, Information Facilities (SCIF) requirements can vary according to the nature of the research, personnel, equipment, and branch of government or institution involved. Each such lab will have specific requirements.

19

Waste Disposal

It is extremely important to be able to predict the kinds and quantities of hazardous waste materials that may be generated in order to design a satisfactory laboratory facility. In operating a laboratory, the janitorial staff should be careful not to mix normal waste products, such as paper, with hazardous chemicals, infectious or radioactive materials.

Chemical Wastes

A dedicated space near the loading dock should be provided to collect and store chemical wastes in preparation for disposal. Such rooms should not be below grade because of the explosion potential. Chemical storage rooms should be on an outside wall containing windows or blow-out panels in a surface to volume ratio of at least one square foot to 40 cubic feet of room volume; refer to NFPA 68. Exhaust ventilation requirements are at least 10 air changes per hour to purge any fumes from spills or leakage. If incompatible chemicals are involved, a physical barrier should subdivide the room. All electrical devices should be explosionproof and a 6A 60BC portable hand fire extinguisher should be mounted on the wall adjacent to the entrance door.

Liquid Wastes

Pathogenic, toxic, and/or acidic liquid wastes require unusual piping systems as do treatment and storage facilities for such wastes—special systems including a central building dilution tank with automatic monitoring to measure the pH of liquids with controls to add caustic or acid as necessary to neutralize the fluids prior to their entry into the municipal sanitary sewer system. In addition, it may be necessary to provide local dilution tanks under certain sinks that will receive heavy concentrations of acid wastes. In extreme cases, it may be necessary to establish a policy for the lab personnel to place such wastes in a reusable container that is periodically disposed of legally offsite.

Radioactive Wastes

Radioactive waste materials must be carefully monitored and captured for appropriate disposal. The Nuclear Regulatory Commission (NRC) issues licenses to qualified scientists for the storage, use, and disposal of radioactive material under very strict guidelines. Facilities that generate radioactive waste require a dedicated, shielded room similar to the chemical storage room near the shipping dock. The NRC also requires continuous air monitoring of exhausts where concentrations are expected to exceed certain predetermined levels. It is important to provide convenient access to these montoring devices in the building design.

Infectious Wastes

Formalin solution is used to kill infectious bacteria during necropsy procedures. It is also a known carcinogen. The sacrificed animals are then kept frozen until disposal. It is standard practice to separate waste materials with the potential for harboring infectious agents from other lab waste materials. The infectious wastes are placed in red colored bags and autoclaved prior to leaving the laboratory. In some large labs onsite incineration is employed, but this is expensive and therefore not a widespread practice. It is also inhibited due to zoning and other governmental regulations. The travel distance within the lab for infectious waste material should be kept to a minimum for obvious safety reasons. Liquid infectious wastes should be treated with a disinfectant "kill" tank prior to disposal into the sewer system.

20

Facilities Management Systems

CADD electronic data provide the ideal base for modern facilities management systems software where alpha-numeric information is combined with building design graphics to provide a comprehensive view of the laboratory system components. Personnel photos and data can be linked to a particular department, telephone extension, funding grant, or workstation. Security alarm detectors can be specifically located with the graphics to indicate a clear picture of potential trouble. Stored hazardous and radioactive material quantities can be kept current and specifically located in the graphics database for a quick overview. Critical pieces of equipment can be located and inventoried with an interactive system that allows real time manipulation. Photographs of people or equipment associated with a particular floor plan location can be linked to the database. Any number of items can be tracked from areas devoted to a funding grant or lease expiration to routine equipment maintenance. *Archibus FM* is one facilities management software that has infinite configuation possibilities in performing these types of tasks. The software can be combined with the direct digital controls (DDC) building and security systems at a PC linked to a shared network. The opportunities are endless for management, security, physical plant, safety, purchasing and personnel departments to oversee the property use and investment.

Since laboratories are frequently occupied 24 hours a day, the critical alarms can be connected via modem to remote locations at anytime. This means that maintenance personnel are not necessarily required onsite whenever the scientists are working during odd hours. Should a crucial exhast fan malfunction at 3 o'clock in the morning, a maintenance person can activate the backup from a remote location, such as his bedside, through the use of this advanced computer-linked technology.

Functions that can benefit from facilities management software include:

- Real Property and Lease Management
- Space and Grant Allocation Management
- Furniture and Equipment Management
- Telecommunications and Cable Management
- Energy Management and DDC Controls
- Materials Stock and Purchasing
- Waste Disposal
- Security, Alarms and Access Control

21

Trends in Lab Design

General

- The lab equipment requirements are becoming more demanding than the needs of the scientists.
- Controlling the laboratory environment and providing reliable utility services has become the highest priority.
- Ultra clean spaces are becoming more common, but with smaller areas.
- Generic bench space is occupying a smaller percentage of lab space.
- Deeper building floorplates due to increased demand for interior spaces remote from influences such as natural light and varying weather.
- Focused teams with shared resources are displacing the individual Principal Investigator with dedicated resources.
- There is more emphasis on interdisciplinary stimulation and open communication of ideas.
- More unusual life forms are being studied.
- There is a greater need for various types of shielding.
- More sophisticated and less complicated forms of security are emerging.
- Lab environments that are adaptable to change provide the best value.
- More dependence on computer modeling and efficient data transmission.
- Highly technical and hazardous spaces will become more isolated.
- More microscale experimentation will take place.
- More dependence on vacuum-cooled environments.
- Greater demand for dedicated grounds, and clean, conditioned power supplies.

- Greater demands for BL-3 and and perhaps BL-4 space as well.
- Less use of optical microscopes and more dependence on other forms of imaging technologies.
- More integration of data, audio, visual and computational capabilities.
- Virtual reality simulation techniques will become more common.
- Use of fiberoptics will become common.
- Security demands will continue to grow.
- More shielding for RF and magnetic interferences will be required.
- More laser equipment will be available requiring vibration dampening
- Focused joint venture research between universities and industry will become more commonplace.

The implications of these trends on lab design are encompassing. Although it is not possible to fully anticipate the impacts of evolving research technologies and processes, there are distinct changes in lab design that must now take place to respond to current and evolving practices.

Equipment

The special services and limited range of acceptable envionmental conditions dictated by a new generation of expensive laboratory equipment and processes are changing the configurations of research facilities. Principal Investigators are becoming more dependent on sophisticated technologies. The use of laser optics, ultra pure fluids, vibration isolation, lighting controls, constant temperature and humidification controls, low temperature storage, various types of imaging and radiation which require shielding for product and personnel protection, exotic pyrophoric and extremely toxic gases, and the ubiquitous ultra clean work space are all contributing to new laboratory design requirements.

The systems, equipment, and services required to maintain these special environments are costly, space-consuming, and due to the rapid pace of technological advancement, subject to rapid change or obsolescence. As a consequence, the enclosing shell and structural components of a research building are less of a priority consideration today than they were just 10 years ago. The increased demands placed on the research building's mechanical, electrical, and process systems that are required to accommodate growing amounts of ultra sensitive equipment have caused this shift. The costs of creating suitable lab space is now driven more by equipment and process needs than by staff amenties. Due to the intense competition for recruiting better scientists, the costs of a modern research facility will be more and more dictated by sophisticated and expensive equipment (in addition to staff amenities and a pleasant work environment).

Priorities

The "world class" scientist is much more interested in the reliability and quality of the electrical power, or the humidification controls, vibration isolation, and the presence of electromagnetic interferences than in his own creature comforts. Of course, it is possible to provide both, but the priorities are clearly in favor of the environmental factors affecting equipment performance. This trend will likely continue. By maintaining clean, sterile environmental conditions, the external influences which can contaminate microbiological studies are reduced. Clean, constant temperature and humidity conditions are often very desirable to maintain both fragile life forms with compromised immune systems and the sound operation of extremely sensitive and delicate equipment capable of measuring parts per billion in sub-molecule components.

Microtome and micro-injection procedures require very low air flow, excellent, glare-free illumination and very clean conditions in the work zone. The levels of cleanliness formerly experienced only in the microelectronics industry are now becoming more commonplace in the biomedical and pharmaceutical arena. Contamination-free work areas with HEPA or ULPA filtration systems are now required in the entire research suite and no longer just within the limits of a biological safety cabinet.

Bench Space vs. Support Space

The ratio of traditional generic lab bench space to support spaces for specialized tasks is changing rapidly, especially in organic chemistry, cell biology, and with recombinant DNA techniques. In a modern clinical laboratory, it is not uncommon for the support space to be at least double the area used for generic benches. This is especially true with the growing trend toward interdisciplinary teams of scientists that focus their combined energies on a specific area of study. Benches are becoming crowded with countertop, computer-controlled automatic equipment to analyze minute details of substances where the technician formerly spent his/her time weighing or counting elements that could be seen with a simple microscope or the unaided eye. The gains in microscope power have been achieved through the use of various specialized devices. These require a customized space with special services, such as chilled water, a variety of lighting conditions, and available electronic grade gases whose purities are measured in parts per trillion. These expensive resources must now be shared among various disciplines if they are to be used economically. This places locational priorities on such equipment within the lab. Ease of access for frequently used equipment is important.

Shared vs. Dedicated Resources

The once popular partitioned dedicated lab module or "turf" assigned to a senior scientist of approximately 11' x 24' containing a fume hood, bench space with sink, gas, vacuum and deionized water, freezers, refrigerators, and other equipment, is rapidly becoming outdated. Its design is not suited for the interdisciplinary teams that are becoming more prevalent in scientific research. It may be more appropriate to have bench areas for larger groups and shared support spaces for better equipment and space utilization. Laboratory floorplates are becoming deeper due to requirements for windowless, controlled artificial environments for much of the lab support zone that contains the specialized equipment.

Idea Exchange

There is more emphasis on the promotion of informal communication and interaction among the researchers, where in the past isolation from distractions was emphasized for concentrated study. The modern lab provides both types of spaces. The group "tea" room concept that originated at Oxford University provides a common small group "oasis" that is conducive to the generation of informal interaction of ideas. Such spaces enhance and tremendously stimulate team oriented research. These informal meeting spaces and the researchers' offices should create a different ambiance from the laboratory which provides variety and improves the quality of the work environment, which in turn is another stimulus to creativity. The oasis concept is predicated upon a casual format of exchange. The room should contain space for journals, makerboards, chairs, tables, and the most essential piece of equipment, the coffee pot. It should be adjacent to, but not directly a part of the oasis.

Basic scientific research was formerly conducted by individuals working alone at the lab bench using small equipment such as the microscope to observe and record data. Today, modern scientists work in collaboration in fluid, interdisciplinary teams. These cross-disciplinary teams frequently occupy different sites and share ideas and data over electronic communication networks. Their interactions are frequent and informal. Compartmentalized work areas by department or Principal Investigator are becoming inappropriate for today's dynamic approach to research.

Animal Types

Rodents, primates, rabbits, and domestic animals have traditionally been used in live tissue studies. With the emergence of gene research, life forms with shorter generational cycles are becoming more popular as biotechnology research subjects. Zebra fish, a popular species found in pet stores for home aquaria, nematodes, a type of invertebrate worm, and insects of all kinds are the current "hot" life forms in gene research. Providing the proper environment for these creatures has added some new challenges for laboratory building systems. The implications of this trend toward smaller species has potentially far-reaching implications on future space requirements and needs of the vivaria and insectories.

Shielding

Most research facilities require some form of built-in shielding to protect personnel, equipment performance, or information. Electromagnetic interference (EMI), nuclear radiation and magnetic energies often must be contained for safe and efficient laboratory working conditions. The available shielding technologies can vary widely in cost and effectiveness. A thorough understanding and definition of acceptable tolerance ranges must be established early in the lab design process for the shielding solution adopted to provide the best value and least impact on the construction cost.

Cross Cultural Partnerships

In the 1980s, federal legislation began to encourage a research alliance between federal laboratories and university programs on one side and U.S. business on the other. Industries can now work to develop new technologies with the assistance of government funding. The Federal Technology Transfer Act of 1986 offered the private sector a degree of patent and intellectual property protection. The General Accounting Office (GAO) had this to say in a 1991 assessment of technology-transfer efforts on behalf of small manufacturers: "The United States has experienced major trade deficits in manufactured goods in each year since 1983, and the U.S. companies have lost significant market share to foreign competitors . . . as several recent studies have noted, the future well-being of the U.S. economy depends on a strong manufacturing base, which requires continual technological improvements to meet growing global competition."

Despite this climate, the potential partners are sometimes reluctant to join resources with such disparate cultural backgrounds. Labs and universities fret that by bending their efforts to satisfy commercial concerns, they may be losing the basic research function that fosters technological innovation in the first place. Businesses complain that the institutional labs, accustomed to the ponderous nature and large budgets of defense related and other noncommercial work, lack the quick response to meet the windows of opportunity in the marketplace. Some business leaders are skeptical of basic scientists. They have the image of being great thinkers, but the word "application" is foreign to them. Nevertheless, with the reduction of defense funding, there will be a greater demand for cooperative research and development agreements between commercial and educational research laboratories.

Adaptability

The cutting edge of science is advancing relentlessly and it is extremely difficult to predict the future courses it will take. Today's expensive modern research facility could become obsolete in a short time. It is imperative for the creators of laboratory facilities to attempt to anticipate change and design adaptable systems and spaces that are capable of adjustment to meet the inevitable differing requirements that the future holds. More than ever before the scientist must be an active participant throughout the programming, financing, planning, and building with a team concept in order to achieve a coherent and integrated laboratory design that will fullfill the need of the present and the future. Although the architect naturally learns from past experience, it is important to approach each new commission with an open mind and without preconceptions. In the dynamic world of scientific research, this is the best mind-set to follow—seek to become better informed and remain flexible!

Appendix

Glossary of Terms–Laboratories

A: Amperes.

AAALAC: American Association for Accreditation of Laboratory Animal Care, a nonprofit corporation formed by leading U.S. scientific and educational organizations to promote high quality animal care and use through a voluntary accreditation program.

AALAS: American Association for Laboratory Animal Science is an organization made up of individuals and institutions professionally concerned with the production, care, and use of laboratory animals.

AAMI: Association for the Advancement of Medical Instrumentation.

Absorption: The process by which a liquid penetrates the solid structure of the absorbent's fibers or particles, which then swell in size to accommodate the liquid.

Access Opening: That part of a fume hood through which work is performed.

ACGIH: American Conference of Governmental Industrial Hygienists. An organization that annually determines standards of exposure to toxic and otherwise harmful materials in the workroom air, commonly known as the TLV.

Acid: A broad category of chemical substances marked, among other things, by sour taste and a propensity to react with alkaline subtances (bases) to form salts.

ACLAM: American College of Laboratory Animal Medicine is a specialty board to encourage education, training and research in animal laboratory medicine.

Action Level: The exposure level (concentration of the material in air) at which certain OSHA regulations to protect employees take effect (CFR 1910.1001-1047); e.g., workplace air analysis, employee training, medical monitoring and record keeping. Exposure at or above the action level is termed occupational exposure.

ADA: The Americans with Disability Act, legislation that prohibits discrimination in four major titles: I: Employment, II: Public Services, III: Public Accommodations and IV: Telecommunications.

Adsorption: The process by which a liquid adheres to the surface of the adsorbent material but does not penetrate the fibers or particles themselves.

Aerosols: Liquid droplets or solid particles dispersed in air, that are of fine enough particle size (.01 to 100 micrometers) to remain so dispersed for a period of time.

AIHA: American Industrial Hygiene Association.

Air Foil: Curved or angular member(s) at the fume hood entrance.

Air Volume: Quantity of air normally expressed in cubic feet per minute (CFM).

Aliphatic: Pertaining to an open-chain carbon compound. Usually applied to petroleum products derived from a paraffin base and having a straight or branched chain, saturated or unsaturated molecular structure. Substances such as methane and ethane are typical aliphatic hydrocarbons.

Alkali, Alkaline: The chemical opposite of acid. Synonym: base. When an alkaline substance reacts with an acid, the two neutralize each other and form a salt.

Allergen: A substance, usually a protein, that causes an allergy.

Alpha Particle: A small electrically charged particle of very high velocity thrown off by many radioactive materials. It is made up of two neutrons and two protons and has a positive electrical charge.

Alpha Ray: A stream of alpha particles.

Alpha Wave: An electrical rhythm of the brain with a frequency of 8 to 13 cycles per second that is often associated with a state of wakeful relaxation.

AMCA: Air Movement and Control Association. Independent group that provides licensed ratings for exhaust fans to ensure credibility of fan performance.

Angstrom: A unit of wavelength of light equal to one ten-billionth of a meter. Ten angstroms equal a nanometer and 10,000 angstroms equal a micron.

Anemometer: A device which measures air velocity. Common types include the rotating vane, the swinging vane, and hot wire anemometer.

ANSI: American National Standards Institute. A private organization that provides the mechanism for creating voluntary standards through consensus. Over 8,000 standards are approved and used widely by industry and commerce.

Antibody: A protein substance made by certain white blood cells in the body in response to injection, ingestion, or inhalation of an antigen.

Antigen: A substance that causes the body's immune system to make a specific antibody that will react with (or neutralize) that antigen.

Apparatus, Laboratory: Furniture, hoods, centrifuges, refrigerators, and commercial or man-made equipment used in a laboratory.

APR: Air-Purifying Respirator.

Aromatic: Applied to a group of hydrocarbons and their derivatives characterized by the presence of the benzene nucleus (molecular ring structure).

Aseptic: Free from pathogenic microorganisms.

ASHRAE: American Society of Heating, Refrigeration and Air.

ASLAP: American Society of Laboratory Animal Practitioners is made up of veterinarians engaged in laboratory animal practice.

Assay: To analyze and quantify a substance.

ASTM: American Society for Testing and Materials. A technical organization which develops standards on characteristics and performance of materials, products, systems, and services. It is the world's largest source of voluntary consensus standards.

atm: Atmospheres.

ATSDR: Agency for Toxic Substances and Disease Registry; federal agency in the Public Health Service charged with carrying out health-related responsibilities of the Comprehensive Environmental Response, Compensation and Liability Act.

Auxiliary Air: Supply or supplemental air delivered to a laboratory fume hood to reduce room air consumption.

AVMA: American Veterinary Medical Association is the major national organization of veterinarians.

Axenic: Free from other living organisms

Background Radiation: The radiation coming from sources other than the radioactive material to be measured. This "background" is primarily due to cosmic rays which constantly bombard the earth's surface from outer space.

Bacteria: General name for a vast variety of microorganisms, including beneficial as well as harmful types.

Bag-in, Bag-out: A system of changing contaminated filters that reduces the potential for personnel contamination.

Bench Mounted Hood: A fume hood that rests on a countertop.

Beta Particle: A small electrical charged particle thrown off by many radioactive materials. It is identical with the electron. Beta particles emerge from the radioactive material at high speeds.

Beta Ray: A stream of beta particles.

Biohazard: An infectious agent, or part thereof, presenting a real or potential risk to the well-being of man, other animals, or plants, directly through infection, or indirectly through disruption of the environment.

Biological Safety Cabinet: (BSC) Special safety enclosure used to handle pathogenic microorganisms (this enclosure is not a fume hood).

Bit or Byte: Basic unit of computer data.

Bit Rate: Speed of computer transmission signal, measured in bits per second, megabits per second, or gigabits per second.

Bonding: A safety practice. The interconnecting of two objects (tanks, cylinders, etc.) with clamps and bare wire. This will equalize the electrical potential between the objects and help prevent static sparks that could ignite flammable materials.

Breakthrough Time: The time from initial chemical contact to detection.

BRI: Building Related Illness.

BSO: Biological Safety Officer.

Bureau of Explosives: Division of the Association of American Railroads that regulates shipping specifications for hazardous products.

Bypass: Compensating opening that maintains a relatively constant volume exhaust through a fume hood, regardless of sash position, and that functions to limit the maximum face velocity as the sash is closed.

CAA: Clean Air Act Public Law PL 91-604; found at 40 CFR 50-80. The regulatory vehicle that sets and monitors airborne pollution that may harm public health or natural resources.

Capture Velocity: The air velocity at the fume hood face, usually expressed in feet per minute (FPM). Sometimes called containment velocity.

Carcinogenic: Cancer causing.

California Hood: A rectangular enclosure used to house distillation apparatus that can provide visibility from all sides with horizontal sliding access doors along the length of the assembly. The enclosure, when connected to an exhaust system, will contain and carry away fumes generated within the enclosure when the doors are closed or when the access opening is limited; (not considered a fume hood).

Canopy Hood: Suspended ventilating device designed to exhaust only heat, water vapor, and odors; (not considered a fume hood).

CAS Number: An assigned number used to identify a material. CAS stands for Chemical Abstracts Service. The numbers have no chemical significance.

CDC: Center for Disease Control.

CFM: Cubic feet per minute.

CFR: Code of Federal Regulations.

CFU: Colony-Forming Units: Microbial growth measurement unit for bioclean rooms.

CGA: Compressed Gas Association.

Characteristic Wastes: Hazardous wastes exhibiting one of four characteristics: ignitability, explosivity, EP toxicity, or corrosivity.

Chelant: A chemical used in stainless steel pipe passivation.

Chromatography: A process of separating, especially a solution of closely related compounds by allowing a solution to seep through an absorbant so that each compound becomes absorbed in a separate, often colored layer.

Chromosomes: The thousands of genes that carry hereditary messages from parents to offspring are strung on 46 microscopic "necklaces" called chromosomes. They are tightly coiled inside most body cells.

Curie (Ci): A unit of radioactivity equal to 0.00000000003.7 disintegrations per second.

Class A Fire: Wood, paper cloth, trash, or other ordinary materials.

Class B Fire: Gasoline, grease, oil, paint, or other flammable liquids.

Class C Fire: Live electrical equipment.

Class D Fire: Flammable metals.

Combination Hood: A fume hood assembly containing a bench hood section and a walk-in section.

Combustible: A term used by NFPA, DOT, and others to classify liquids that will burn on a basis of flash points. Both NFPA and DOT generally define combustible liquids as having a flash point of 100°F (38°C) or higher. Nonliquid substances such as wood and paper are classified as ordinary combustibles by NFPA. OSHA defines combustible liquid within the Hazardous Communication Law as any liquid having a flash point at or above 100°F, but below 200°F.

Containment: The term used in describing safe methods for managing infectious agents in the laboratory environment where they are being handled or maintained. The three elements of containment include laboratory practice and technique, safety equipment, and facility design.

Corrosive: A substance that causes visible destruction or permanent changes in human skin tissues at the site of contact. Capable of destroying tissue, as a strong acid or alkali.

Cross Draft: A flow of air that blows into or across the face of a fume hood.

Cryogen: A substance for obtaining very low temperatures.

Cryogenic Fluid: Substance that exists only in the vapor phase above –73°C at one atmosphere pressure and that is handled, stored, and used in the liquid state at temperatures at or below –73°C while at any pressure.

Cryogenics: A branch of physics that deals with the production and effects of very low temperatures.

Culture: A method of growing cells or microorganisms in the laboratory in order to identify them and to determine their resistance to antibiotics. The special food on which they are grown is the culture medium.

CWA: Clean Water Act; Public Law PH 92-500, found at 40 CFR 100-140 and 400-470. The EPA and the Corps of Engineers have jurisdiction. CWA regulates the discharge of nontoxic and toxic pollutants into surface waters.

Cyclotron: A device that can prepare a number of different radioactive solutions on demand. It does so by combining substances such as glucose with radioactive tags such as the positron emitting isotopes of oxygen, nitrogen, fluorine, or carbon.

Damper: A device installed in a duct to control air flow volume.

dB: Decibel. A unit for expressing the relative intensity of sounds on a scale of 0 (average least perceptible) to 130 (average pain level).

DDC: Direct Digital Control.

Demonstration Hood: An enclosure used for student demonstrations that has visibility and, normally, access from both sides. Used primarily in schools for control of noxious fumes (not considered a laboratory fume hood).

Dielectric: A material which is an electrical insulator on which an electric field can be sustained with a minimum dissipation of power.

Diffraction: A modification which light undergoes in passing by the edges of opaque bodies or through narrow slits or in being reflected from ruled surfaces and in which the rays appear to be deflected and to produce fringes of parallel light and dark or colored bands. Also applies to other waves, as in sound.

Distillation Hood: A laboratory fume hood that provides a work surface approximately 18″ above the floor, to accommodate tall apparatus.

DNA: Deoxyribonucleic acid; the building blocks of life. It carries life's basic genetic code. Recombinant DNA techniques are man-made changes in the genetic material of living organisms. Using gene-splicing techniques, scientists have already created organisms to treat disease, improve agriculture, and even gobble oil spills.

Drop Test: A test required by DOT regulations for determination of the quality of a container or finished product.

Dry Bulb Temperature: The temperature of air measured with a dry bulb thermometer in a psychrometer to measure relative humidity.

Duct: Round, square, oval, or rectangular tube used to enclose moving air.

Duct Velocity: Speed of air moving in a duct, usually expressed in feet per minute (FPM).

Dusts: Solid particles generated by handling, crushing, grinding, rapid impact, detonation, and decrepitation of organic or inorganic materials, such as rock, ore, metal, coal, wood, and grain. Dusts do not tend to flocculate, except under electrostatic forces; they do not diffuse in air, but settle under the influence of gravity.

Electroencephalograph, (EEG): A machine that helps neurologists detect electronic waves emanating from different areas of the brain.

Endocrine System: An interlocking directorate of glands whose hormones control bodily growth, sex characteristics, metabolism, and many other functions.

Enzymes: Proteins made by the body that act as catalysts for many biochemical reactions.

EP Toxicity: Extraction-procedure; toxicity tests performed on RCRA wastes.

EPA: U.S. Environmental Protection Agency. A federal agency with environmental protection regulatory and enforcement authority.

Etiologic Agent: Organisms, substances, or objects associated with cause of disease or injury.

Exhaust Ventilation: The removal of air, usually by mechanical means from any space. The flow of air between two points is due to the occurrence of a pressure difference between the two points. This pressure difference will cause air to flow from the high pressure to the low pressure zone.

Explosion Class 1: Flammable gas or vapor.

Explosion Class 2: Combustible dust.

Explosion Class 3: Ignitable fibers.

Explosion-Proof: An electrical apparatus so designed that an explosion of flammable gas or vapor inside the enclosure will not ignite flammable gas or vapor outside.

Exposure Level: The level or concentration of a physical or chemical hazard to which an employee is exposed.

Exposure Limits: Concentration of substances (and conditions) under which it is believed that nearly all workers may be repeatedly exposed day after day without adverse effects. ACGIH limits are called TLV and OSHA exposed limits are called PEL (permissible exposure limits).

FA: Fresh Air.

Face Velocity: Average air velocity into the exhaust system measured at the opening into the hood or booth.

FDA: United States Food and Drug Administration.

FDDI: Fiber Distributed Digital Interface, a standard in data networking on a fiber optic medium using token ring methodology consisting of two counter rotating rings.

FEMA: Federal Emergency Management Agency; body responsible for administering certain training funds under SARA Title III.

FEMA: Fire Equipment Manufacturers Association.

Femtotesla (fT): An extremely small signal of magnetic force, used to measure magnetic fields emanating from the brain. Much smaller than one-billionth of the earth's magnetic force.

Flame Arrestor: A mesh or perforated metal insert within a flammable storage can which protects its contents from external flame or ignition.

Flammable: Flash point less than 100°F (and a vapor pressure of not over 50 psia at 100°F) (definition may vary by organization).

Flammable Liquid: A liquid with a flash point below 100°F (37.8°C), excluding gases.

Flash Back: The phenomenon characterized by vapor ignition and flame travel back to the vapor source (the flammable liquid).

Flash Point: The lowest temperature at which a flammable vapor-air mixture above the liquid will ignite when the ignition source is introduced.

FM: Factory Mutual. A nationally recognized testing laboratory and approved service recognized by OSHA.

Formalin: A clear aqueous solution of formaldehyde containing a small amount of methanol, used in postmortem examine work.

Frazier Air Test: A testing method of the ASTM for material breathability.

FRP: Fiberglass reinforced polyester.

ft-c: Footcandles, the amount of illumination cast on a surface one foot from a lighted candle.

Fume Removal System: A combination of laboratory fume hood, make-up or supply air, auxiliary air (if used), exhaust ducts, exhaust fan, and pollution abating device (if used).

Fumes: Particulate matter consisting of the solid particles generated by condensation from the gaseous state, generally after volatilization from melted substances, and often accompanied by a chemical reaction, such as oxidation.

Fungus: A parasitic microorganism. Many are villains, but some produce antibiotics.

Gamma-Rays: The most penetrating of all radiation. High-energy protons, especially as emitted by a nucleus in a transition between two energy levels.

Gas: A state of matter in which the material has very low density and viscosity; can expand and contract greatly in response to changes in temperature and pressure; easily diffuses into other gases; readily and uniformly distributes itself throughout any container.

Gauss: A unit of measurement of a magnetic field. The earth's magnetism is about 0.5 gauss. 10,000 gauss equals one tesla.

GBPS: Gigabits per second.

Gel: A substance of jelly-like consistency.

Genetic: Inherited through the parents' genes.

Germicidal: Referring to chemical agents—lethal to germs— which are generally used on inanimate objects.

GFI: Ground fault circuit interrupter.

Gigabit (GB): 1,000,000 bits or bytes.

Glove Box: Total enclosure used to confine and contain hazardous materials with operator access by means of gloved portals or other limited openings; (not a laboratory fume hood).

GLP: Good Laboratory Practices.

GMP: Good Manufacturing Practices as established by the Food and Drug Administration.

Gnotobiotic: Of, relating to, living in, or being a controlled environment containing one or a few kinds of organisms.

Gram: A unit of weight in the metric system. Approximately 28 grams equals one ounce.

Hazard Classes: A series of nine descriptive terms that have been established by the United Nations Committee of Experts to categorize the hazardous nature of chemical, physical, and biological materials. These categories are: flammable liquids, explosives, gases, oxidizers, radioactive materials, corrosives, flammable solids, poisonous and infectious substances, and dangerous substances.

Hazardous Material (DOT): A substance or material which has been determined by the Secretary of Transportation to be capable of posing an unreasonable risk to health, safety, and property when transported in commerce and so designated (49 CFR 171.8).

Hazardous Waste: Under RCRA, any solid or combination of solid wastes, which because of its physical, chemical, or infectious characteristics, may pose a hazard when improperly managed.

HCS: Hazard Communication Standard; the OSHA standard cited in 29 CFR 1910.1200 requiring communication of risks from hazardous substances to workers in regulated facilities.

HEPA: High Efficiency Particulate Absolute. A filter which removes from air 99.97% or more of monodisperse dioctyl phthalate (DOP) particles having a mean particle diameter of 0.3 micrometer. Common use—"HEPA filter" high efficiency particulate air filter.

HMAC: Hazardous Materials Advisory Council; national organization representing the hazardous materials industry. Members are devoted to domestic and international safety in transportation, and handling of hazardous materials and waste.

HMIG: Hazardous Materials Identification Guide.

HMR: Hazardous Materials Regulations; regulations administered and enforced by various agencies of DOT governing the transportation of hazardous materials by air, highway, rail, water, and intermodal means.

Hormone: A chemical substance made in an endocrine gland and secreted into the bloodstream. The hormone then acts on some distant target within the body.

HPM: Hazardous Production Materials (liquids, solids, and gases) listed in the Uniform Building Code and the Uniform Fire Code for Group H, Division 6 occupancy; usually associated with wafer fabrication facilities.

HSWA: Hazardous and Solid Waste Amendments; 1984 amendments to RCRA establishing a timetable for RCRA land bans and more stringent requirements for RCRA activities.

HVAC: Heating, Ventilating, and Air Conditioning.

IBC: Institutional Biosafety Committee.

IDLH: Immediately Dangerous to Life and Health; maximum concentration of a chemical in air to which one can be exposed without suffering irreversible health effects (function of time, usually).

Inches of Water Gauge (in w.g.): A unit of pressure equal to the weight of a column of liquid water one inch high at 20°C (1 inch w.g. = 0.036 psi).

Incubate: To facilitate growth or multipication of cells or germs in a culture medium (in vitro) to the point where they can be identified, or in vivo to the point where they cause symptoms.

Indoor Air Quality (IAQ): Sick building syndrome, tight building syndrome. The study, evaluation, and control of indoor air quality related to temperature, humidity, and airborne contaminants.

Infectious: Capable of invading a susceptible host, replicating, and causing an altered host reaction, commonly referred to as disease.

Infectious Agent: Agents capable of producing a disease or abnormal response in man, animals, or a tissue culture system.

In Vitro: Literally, "in glass"—a medical or biological event that takes place, outside the living body, in the laboratory. As opposed to in vivo.

In Vivo: Literally, "in life"—a medical or biological event that takes place in a living human or animal. As opposed to in vitro.

Ionizing Radiation: High energy radiation, such as that produced by X-rays, gamma rays from radio isotopes, and nuclear fallout, which penetrate deep into bodily tissue.

ISDN: Integrated Services Digital Network.

Isotonic: Having the same osmotic pressure as the fluid phase of a cell or tissue.

Joule: Unit of energy used in describing a single pulsed output of a laser. It is equal to one wattsecond or 0.239 calories.

kg: Kilogram.

Kinesiology: The study of human movement in terms of functional anatomy.

Lab Pack: Generally refers to any small container of hazardous waste in an overpacked drum, but not restricted to laboratory wastes.

Laminar Air Flow: Air flow in which the entire body of air within a designated space moves with uniform velocity along parallel flow lines.

Laminar Flow Cabinet: Name applied to clean bench or biological safety cabinet that incorporates a smooth directional flow of air to capture and carry away airborne particulates.

LAN: Local Area Network.

Laser: Light amplification by Stimulated Emission of Radiation.

LC: Lethal Concentration.

LCD: Liquid Crystal Display. A constantly operating display that consists of segments of a liquid crystal whose reflectivity varies according to the voltage applied to them.

LD: Lethal Dose.

Leakage Radiation: X-rays emitted through a protective housing of the X-ray tube in directions other than the useful beam.

LED: Light-Emitting Diode. A semiconductor diode that converts electric energy efficiently into spontaneous and noncoherent electromagnetic radiation at visible and near-infrared wavelengths.

LEL : Lower Explosive Limit. The maximum percent by volume of a gas which, when mixed with air at normal temperature and pressure, will form a flammable mixture.

LEPC: Local Emergency Planning Committee; groups defined in the Superfund Amendment and Reauthorization Act (SARA), a federal law reauthorizing and expanding the jurisdiction of the Comprehensive Environment Response, Compensation and Liability Act of 1980 (CERCLA).

Level A Clothing: Should be worn when highest level of respiratory, skin, and eye protection is needed.

Level B Clothing: Should be worn when highest level of respiratory protection is needed, but a lesser level of skin protection.

Level C Clothing: Should be worn when the criteria for using air-purifying respirators are met.

Level D Clothing: Should be worn only as a work uniform and not on any site with respiratory or skin hazards.

Liner: Interior lining for a laboratory fume hood, including sides, back, top, exhaust plenum, and baffle.

Local Exhaust Ventilation: A ventilation system that captures and removes the contaminants at the point they are being produced before they escape into the workroom air.

LSM: *Laboratory Safety Monograph*, available from the National Institutes of Health Office of Recombinant DNA Activities

Make-Up Air: Air needed to replace the air taken from the room by laboratory fume hoods and other exhausts.

Manometer: Device used to measure air pressure differential, usually calibrated in inches of water.

Maximum Use Concentration (MUC): The product of the protection factor of the respiratory protection equipment and the permissible exposure limit (PEL).

MBPS: Megabits per second.

Medium: What bacteria feed on—the special mixture of nutrients, chemicals, and fluids in which and on which cultured cells and microoganisms grow in laboratories.

MEG: Magneto Encephalogram, an imaging procedure using a bio-magnetometer instrument to measure the strength and location of natural bioelectric currents within the body.

Megabit: 100,000 bits or bytes.

MG: Motor Generator.

Micron: A unit of microscopic measurement convenient for describing cellular dimensions or electronic path width on semi-conductor chip; 1/1,000 of a millimeter, or 1/25,000 inch, abbreviated μ.

Microorganisms: Living creatures too small to be seen with the naked eye. They include bacteria, viruses, and fungi.

mil: one mil equals 1/1000 of an inch.

Milligram: 1/1,000th of a gram; a unit of weight in the metric system.

Mists: Suspended liquid droplets generated by condensation from the gaseous to the liquid state or by breaking up a liquid into a dispersed state, such as by splashing, foaming, or atomizing. Mist is formed when a finely divided liquid is suspended in air.

Monodisperse Aerosol: An areosol containing particles of nearly the same size.

MPPCF: Million particles per cubic foot.

MRI: Magnetic Resonance Imaging, a noninvasive imaging technology that uses a strong magnetic field and a pulsating radio frequency to excite hydrogen atoms in the body.

MSI: Magnetic Source Imaging, a noninvasive imaging technique that measures the location and strength of bioelectric currents within the body.

Mutagen: A substance that tends to increase the frequency or extent of mutations (mustard gas, various radiations).

mW: Milliwatt

Nanometer: A very small metric unit of measure equal to 1/1,000th of a micron, or one-billionth of a meter

NAS: National Academy of Science.

NBS: National Bureau of Standards [now the National Institute of Standards and Technology].

NCI: National Cancer Institute.

Negative Air Pressure: Air pressure lower than ambient. Pressure in a space that causes an inflow of air.

Necropsy: Postmortem examination.

Nephrology: A branch of medical science dealing with the kidney—its structure, functions, and diseases.

NFPA Hazard Rating: Classification of a chemical by a four- color diamond representing health, flammability, reactivity and special notice by a numbered hazard rating from 0 to 4.

NFPA: National Fire Protection Association. An organization which promotes knowledge of fire protection methods.

ng: Nanogram.

NIOSH: National Institute of Occupational Safety and Health; independent federal agency charged with performing research on occupational disease and injury.

N.O.: Normally Open.

NOAA: National Oceanic and Atmospheric Administration; scientific support organization serving regulatory agencies charged with enforcing environmental laws affecting oceans and atmosphere.

NRC: Nuclear Regulatory Commission.

NRR: Noise Reduction Rating.

NSF: National Sanitation Foundation International, an independent nonprofit organization that serves the consumer, government, and industry in developing solutions for problems pertaining to public health and the environment.

NSF: National Science Foundation.

NTP: National Toxicology Program.

NTP: Normal Temperature and Pressure, which is defined as 70°F and 14.696 psia.

OA: Ouside Air.

Odor Threshold: The minimum concentration of a substance at which a majority of test subjects can detect and identify the characteristic odor of a substance.

Oncology: A branch of medical science dealing with tumors (particularly cancers)—their origin, nature, growth, and treatment.

Optical Density (OD): A logarithmic expression of the attenuation afforded by a filter.

ORDA: NIH Office of Recombinant DNA Activities.

OSHA: Occupational Safety and Health Administration; oversees and regulates workplace health and safety.

PAPR: Powered air purifying respirator.

Partial Containment Enclosure: An enclosure which is constructed so that contamination between its interior and the surroundings is minimized by the controlled movement of air. Class I and Class II biological safety cabinets are an example.

Passivation: A procedure for treating stainless steel piping with nitric acids or polyfunctional organic carboxylic acids, or chelants, to remove potentially corrosive impurities, contaminants, and rustable surface free iron. A thin coating of relatively unreactive surface film is thus created making the pipe suitable for ultra pure fluids such as water for injection (WIF).

Pathological: Describing an abnormality usually caused by disease.

PCB: Polychlorinated biphenyl. A pathogenic and teratogenic industrial compound used as a heat transfer agent; PCBs may accumulate in human or animal tissue.

PDU: Power Distribution Unit.

PEL: Permissible Exposure Limit. The OSHA limit of employee exposure to chemicals; found primarily in 29 CFR 1910.1000.

Perfusion: To force a liquid through an organ or tissue, especially by way of the blood vessels.

Permeation Rate: An invisible process by which a hazardous chemical moves through a protective material.

PET: Positron Emission Tomography, an imaging technology that can help determine abnormalities in living tissue after a radioactive isotope has been injected into the blood stream.

pH: Means used to express the degree of acidity.

Pharmacological: Concerning the action—therapeutic or toxic or both—of a drug on the body; its absorption, metabolism, and excretion, as well as its effect on cells, tissues, organs, and bodily functions.

Phytates: Chemical substances, salts of phytic acid, which have the capacity to combine with calcium and iron, thus impairing the body's absorption of these nutrients.

P.I.: Principal Investigator.

Plenum: An enclosure for flowing gases in which the static pressure at all points is relatively uniform.

PMA: Pharmaceutical Manufacturing Association.

Poison A: Poisonous gases or liquids of such toxicity that a very small amount of the gas or vapor mixed with air is dangerous to life.

Poison B: Substances, liquid or solid, other than Class A poisons or irritating material which are known to be so toxic to humans as to afford a hazard to health during transportation, or which, in the absence of adequate data on human toxicity are presumed to be toxic to man based on prescribed tests on laboratory animals.

Positive Air Pressure: Air pressure higher than ambient; pressure in a space that causes an outflow of air.

POU: Point of Use.

PPE: Personal Protective Equipment. Devices worn by the worker to protect against hazards in the environment. For example, respirators, gloves, and hearing protectors.

ppm: Parts per million. A convenient means of expressing very low concentrations of a substance in a mixture, or as a low level contaminant in a pure product.

Prefilter: A filter used in conjunction with a cartridge on an air-purifying respirator. A filter placed in the air stream prior to the primary filter.

psi: pounds per square inch, i.e., pressure.

psig: pounds per square inch gauge.

Pulmonary: Having to do with the lungs.

PVC: Polyvinylchloride. A member of the family of vinyl resins.

RAC: Recombinant DNA Advisory Committee, a 25-member group that advises the Secretary of Health and Human Services and the Director of the National Institutes of Health.

Radioactive Isotope: A kind of chemical element, either natural or made in a nuclear reactor, that emits energy in the form of radiation that can be detected by instruments.

Radionuclide: A radioactive nuclide; one that has the capability of spontaneously emitting radiation.

RCRA: Resource Conservation and Recovery Act. Regulates materials and wastes currently being generated, treated, stored, disposed, or distributed.

Reactivity: A substance's susceptibility to undergoing a chemical reaction or change that may result in dangerous side effects, such as an explosion, burning, and corrosive or toxic emissions.

Real-Time Sampling: The sample time interval is as short as possible and the acquired data closely approximates the signal.

Recombinant DNA Molecules: Viable organisms containing molecules made outside living cells by joining natural or synthetic DNA segments to DNA molecules that can replicate in a living cell, or DNA molecules that can result from the replication of those described above.

Regulated Material: A substance or material that is subject to regulations set forth by the Environmental Protection Agency (EPA), the Department of Transportation (DOT), or any other governmental agency.

Relative Humidity: The ratio of the quantity of water vapor present in air to the quantity that would saturate it at any specific temperature.

Repetitive Sampling: Samples are taken when a variable, but known time interval has occurred after a trigger event.

Reverse Osmosis Water Treatment Unit: A device installed on a potable water system through which water flows for the reduction of total dissolved solids concentration using the reverse osmosis method.

RF: Radio Frequency.

RFI: Radio Frequency Interference.

Rheumatology: A branch of medical science dealing with rheumatic diseases (such as rheumatoid arthritis and gout) or diseases of connective tissue.

R.O.D.I.: Reverse osmosis, deionized water.

RSO: Radiation Safety Officer, the individual responsible for establishing and maintaining an adequate radiation control program for a research facility, usually a medical physicist.

Safety Equipment: Includes biological safety cabinets and a variety of enclosed containers; see "Containment."

SAMA: Scientific Apparatus Makers' Association.

Sanitize: To clean, not to sterilize.

SARA: Superfund Amendment and Reauthorization Act. Federal law reauthorizing and expanding the jurisdiction of CERCLA.

SARA Title III: Part of SARA mandating public disclosure of chemical information and development of emergency response plans.

SBS: Sick Building Syndrome.

Scatter Radiation: Usually X-rays scattered from the target subject and from adjacent objects.

SCBA: Self-contained breathing apparatus. Designed for entry into and escape from atmospheres Immediately Dangerous to Life or Health (IDLH) or oxygen deficient.

SCIF: Secure, compartmented, information facility. Generally considered a Department of Defense term for secret research space.

Solute: A dissolved substance.

Solution: Mixture in which the components lose their identities and are uniformily dispersed. All solutions are composed of a solvent and the substance dissolved, called the solute.

SPECT: Single Photon Emission Computed Tomography.

Spectrum: A range of frequencies within which radiation has some specified characteristic, such as audio frequency spectrum, ultraviolet spectrum or radio spectrum.

SQUID: Superconducting Quantum Interference Device, a sensing coil that works at extremely low temperatures within a biomagnetometer.

Static Pressure: The potential pressure exerted in all directions by a fluid at rest. When added to velocity pressure, it gives total pressure. Usually expressed in inches of water. In ventilation applications, static pressure is usually the difference between the absolute pressure in an exhaust system and atmospheric pressure.

Static Pressure Loss: Measurement of resistance created when air moves through a duct or hood; usually expressed in inches of water.

STEL: Short-Term Exposure Limit; maximum concentration for a continuous 15 minute exposure period. (Maximum of four such periods per day, 60 minutes minimum between exposure periods, and the daily TLV-TWA must not be exceeded.)

Sterile: Free of living microorganisms. The absence of life on or in an object. This is an absolute term; there can be no such description as "nearly sterile," "partially sterile," etc.

Sterilize: Any process, physical or chemical, which results in the absence of all life or in an object, applied especially to microorganisms, including bacteria, fungi, and their spores, and the inactivation of viruses.

Supplied Air: Breathable air supplied to a worker's mask/hood from a source outside the contaminated area.

Table Top Hood: A small, spot ventilation hood for mounting on a table that is normally vented down through the table top. Used primarily in educational laboratories to control noxious fumes (not considered a laboratory fume hood).

TEE: Transesophageal Echocardiograph. A type of diognostic equipment that provides a three-dimensional image of a working heart.

Teratogenic: Tending to cause development malformations and monstrosities.

Tesla (T): A unit of measurement for a magnetic field equal to 10,000 gauss. The earth's magnetic field measures approximately 0.5 gauss.

Threshold Limit Value (TLV): An estimate of the average safe airborne concentration of a substance and representing conditions under which it is believed that nearly all workers may be repeatedly exposed day after day without adverse effect. The TLV values are published annually by the American Conference of Governmental Industrial Hygienists (ACGIH) in their TLV Booklet. TLV is a trademark of ACGIH.

Tier I or **Tier II**: Inventory form required under SARA Title III for reporting quantities and locations of hazardous substances.

Time Weighted Average (TWA): Usually, a personal 8 hour average exposure concentration to an airborne, chemical hazard.

TNT: Trinitrotoluene.

Tomography: A diagnostic technique using X-ray photographs in which the shadows of structures before and behind the section under scrutiny do not show.

Transport Velocity: Minimum speed of air required to support and carry particles in an air stream.

TVSS: Transient Voltage Surge Suppressor.

UFC: Uniform Fire Code.

UHP: Ultra High Purity, usually with regard to gas distribution systems.

UL: Underwriters Laboratories. An independent, nonprofit organization which operates laboratories for the investigation of devices and materials in respect to hazards affecting life and property.

Ultraviolet (UV): Wavelengths of the electromagnetic spectrum which are shorter than those of visible light and longer than X-rays.

UPS: Uninterruptable Power Source.

Urology: A branch of surgery dealing with the urinary tract—kidneys, ureters, bladder, urethra—plus (in males) the prostate gland and genitals.

USP: United States Pharmacopeia, a semiofficial pharmacological directory of drug standards and specifications.

Vapors: The gaseous form of substances that are normally in the solid or liquid state (at room temperature and pressure).

VAV: Variable Air Volume.

Velocity Pressure: Pressure caused by moving air in a laboratory fume hood or duct, usually expressed in inches of water.

Viable: Literally, "capable of life." Generally refers to the ability of microbial cells to grow and multiply as evidenced by formation of colonies on an agar culture medium; or, as with viruses, to divert the host cell's metabolism to replications of the parasite.

Virus: A term for a group of microbes which, with few exceptions, are capable of passing through fine filters that retain bacteria; they are incapable of growth or reproduction apart from living cells.

Viscosity: The property of a fluid that resists internal flow by releasing counteracting forces.

VOC: Volatile Organic Compound.

Walk-In Hood: A floor-mounted, full height fume hood, designed to accommodate tall apparatus and permit roll-in of instruments and equipment.

Wavelength: The distance in the line of advance of a wave from any point to a like point on the next wave. It is usually measured in angstroms, microns, micrometers, or nanometers.

X-Ray: Highly penetrating radiation. Unlike gamma-rays, X-rays do not come from the nucleus of the atom, but from the surrounding electrons.

Laboratory Office Survey

Laboratory	Staff Title	SF/Person	Type
AT & T Laboratories	Senior Staff	150	Private
New Jersey	Staff	75	Shared
Applied Physics Lab Johns Hopkins Maryland	Staff	75	Shared
IBM Almeden Laboratories	Staff	120	Private
California	Assistant Staff	60	Shared
Hewlett-Packard California	Staff	120	Private
Genelabs	Staff	100	Private
California	Assistant Staff	50	Open_
Genetech	Manager	120	Private
California	Senior Professional	120	Private
	Group Leader	85	
	Technician	80	Shared
Allied Signal Lab	Manager	150	Private
New Jersey	Senior Professional	100	Private
American Cyanamid	Manager	150	Private
Connecticut	Group Leader	120	Private
	Senior Scientist	70	
	Staff Scientist	50	
AT & T Bell Labs Pennsylvania	Staff	75	Shared
Bausch & Lomb	Manager	150	Private
New York	Professional	120	Private
	Junior Professional	60	
	Technician	40	

Laboratory	Staff Title	SF/Person	Type
Eli Lilly Research Center Indiana	Scientist	80	-
Miles Laboratories Connecticut	Manager	125	Private
	Surgical Scientist	100	Private
	Lab Technician	20	Open
Raytheon Lab Massachusetts	Manager	160	Private
	Engineer	80	-
	Technician	64	-
Sandia Laboratories New Mexico	Division Supervisor	140	Private
	Staff	100	Private
	Staff	70	Shared
Sterling Laboratories Pennsylvania	Senior Scientist	120	Private
	Scientist	72	-
Travenol Laboratories Illinois	Senior Researcher	130	Private
	Research Assistant	54	-
MGH Martin Labs Massachusetts	Principal Investigator	150	Private
MIT Lincoln Lab Massachusetts	Group Leader	160	Private
	Senior Staff	160	Private
	Staff	120	Shared
	Assistant Staff	80	Shared

Metrics

Unlike the construction industry in the United States, researchers generally use the metric system in their work. It is easier for scientists to communicate measurements using metric units, so it follows that architects who deal with the scientific community need to thoroughly understand the metric measurements. Besides, it is inevitable that even the construction community will soon be dimensioning in meters and millimeters rather than feet and inches.

Basic Units:
METER : a little longer than a yard (about 1.1 yards)
LITER: a little larger than a quart (about 1.06 quarts)
GRAM: a little more than the weight of a paper clip

Common Prefixes:
milli: one-thousandth (0.001)
centi: one-hundredth (0.01)
kilo: one-thousand times (1,000)

For Example:
1,000 millimeters = one meter
100 centimeters = one meter
1,000 meters = one kilometer

Other Commonly Used Units:
millimeter: 0.001 meter: diameter of a paper clip wire
centimeter: 0.01 meter: a little more than the width of a
    paper clip (about 0.4 inch)

kilometer: 1,000 meters: somewhat further than 1/2 mile
(about 0.6 mile)
milliliter: 0.001 liter: five of them make a teaspoon
hectare: about 2.5 acres
metric ton: about one ton

There are seven (7) metric base units of measurement.

Quantity	Unit (symbol)
length:	meter (m)
mass: ("weight")	kilogram (kg)
time:	second (s)
electric current:	ampere (A)
temperature*:	kelvin (K)
luminous intensity:	candela (cd)
mole:	molecular substance

* Celsius temperature (° C) is more commonly used than kelvin, but
both have the same temperature gradients. Celsius temperature is
simply 273.15 degrees warmer than kelvin, which begins at absolute zero. For instance, water freezes at 273.15 K and at 0°C; it
boils at 373.15 K and at 100°C. To move between Celsius and
kelvin, add or subtract 273.15.

Length: *
1 inch=25.4 millimeters
1 millimeter =0.03937 inch
1 foot=0.3048 meter
1 yard =0.9144 meter
1 meter=3.281 feet

* Use only the meter and millimeter in building design and construction. Avoid use of the centimeter.

Area:
1 square foot =144 square inches
1 square foot=0.0929 square meter
1 square meter=10.76 square feet
1 acre=0.4047 hectare (ha)
1 hectare=2.471 acres

Volume:

1 cubic inch=16.39 cubic centimeters
1 cubic cm=0.06102 cubic inch
1 cubic foot=1728 cubic inches
1 cubic yard=0.7648 cubic meter
1 cubic meter=1.308 cubic yard

Visualizing Metrics:

- One millimeter (mm) is about 1/25 inch, or slightly less than the thickness of a dime.
- One meter (m) is the length of a yardstick plus about 3.33 inches.
- One gram (g) is about the mass (weight) of a large paperclip.
- One kilogram (kg) is about the mass (weight) of a soft-bound model building code book (2.2 lb).
- One liter (L) is about the volume of a 4 inch cube—a little over a quart.
- One liter of water has a mass of 1 kilogram.
- One inch is just a fraction (1/64 inch) longer than 25 mm (25 mm = 63/64 inch).
- Four inches are about 1/16 inch longer than 100 mm.
- One foot is about 3/16 inch longer than 300 mm.
- Four feet are about 3/4 inch longer than 1200 mm.

Metric Drawing Scales

Metric scales are true ratios and are the same for both architectural and engineering drawings. Preferred scales are:

1:1	Same as full size
1:5	Close to 3" = 1'-0"
1:10	Between 1" = 1'-0" and 1-1/2" = 1'-0"
1:20	Between 1/2" = 1'-0" and 3/4" = 1'-0"
1:50	Close to 1/4" = 1'-0"
1:100	Close to 1/8" = 1'-0"
1:200	Close to 1/16" = 1'-0"
1:500	Close to 1" = 40'
1:1000	Close to 1" = 80'

Length Conversion:

Convert from	To	Multiply by
inch	millimeter (mm)	25.4 *
inch	centimeter (cm)	2.54 *
foot	meter (m)	0.3048 *
yard	meter (m)	0.9144 *
chain	meter (m)	20.1168
mile (statute)	kilometer (km)	1.609347
mile (nautical)	Kilometer (km)	1.852 *
millimeter (mm)	inch	0.039370
centimeter (cm)	foot	0.032808
meter (m)	foot	3.280840
meter (m)	yard	1.093613
kilometer (km)	mile (statute)	0.621370

* Indicates exact value

Area Conversion:

Convert from	To	Multiply by
square inch	square centimeter	6.4516 *
square foot	square meter	0.092903
square yard	square meter	0.836127
square mile	square kilometer	2.589998
acre	hectare	0.404687
square centimeter	square inch	0.155000
square meter	square foot	10.76391
square meter	square yard	1.19599
square kilometer	square mile	0.386101
hectare	acre	2.471044

* Indicates exact value

Bibliography

Ashbrook, P. C., and M. M. Renfrew, *Safe Laboratories: Principles and Practices for Design and Remodeling,* Lewis Publishers, Inc., Chelsea, MI, 1990.

ASHP Technical Assistance Bulletin on Handling Cytotoxic and Hazardous Drugs. *American Journal of Hospital Pharmacy,* 47, 1033-104 (1990).

ASHRAE HVAC Applications Handbook, 1991 Edition, Atlanta, GA.

Biosafety in Microbiological and Biomedical Laboratories, U.S. Department of Health and Human Services, U.S. Government Printing Office, Washington, DC.

Doors and Hardware; Door and Hardware Institute, July, 1993.

Guide for the Care and Use of Laboratory Animals; NIH Publication No. 86-23, Revised 1985.

Guidelines for Laboratory Safety: Health and Safety Considerations, John Wiley & Sons, New York, NY.

Guidelines for the Laboratory Use of Chemical Substances Posing a Potential Occupational Carcinogenic Risk; Laboratory Chemical Carcinogen Safety Standards Subcommittee of the DHEW Committee to Coordinate Toxicology and Related Programs.

Harvard Guidelines for Chemical Fume Hood; Appendix D, B3.2(A), Chemical Fume Hood.

IES Recommended Practices: IES-RP-CC-002-86: Laminar Flow Clean Air Devices. Mount Prospect, IL, Institute of Environmental Sciences, January 1986.

Laboratory Fume Hood Standards; Recommended for the U.S. Environmental Protection Agency. Contact No. 68-01-4661. January 15, 1972.

Laboratory Fume Hoods; Laboratory Apparatus Section, Scientific Apparatus Makers Association, SAMA Standard LF-10-1980.

Laboratory Safety Monograph; A supplement to the NIH Guidelines for Recombinant DNA Research. DHEW, Public Health Service, NIH, January, 1979.

Laboratory Safety: Principles and Practices, Miller, B. M., Ed. American Society for Microbiology, Washington, DC, 1986.

McDermott, H. J., *Handbook of Ventilation for Contaminant Control;* Ann Arbor Science Publishers, Ann Arbor, MI.

Medical Laboratory Planning and Design, College of American Pathologists, Skokie, IL

Minimum Acceptable Face Velocities of Laboratory Fume Hoods and Guidelines for their Classification; Oak Ridge National Laboratory #ORNL/TM 7400, Oak Ridge, TN.

National Sanitation Foundation Standard Number 49 for Class II (Laminar Flow) Biohazard Cabinetry, Ann Arbor, MI, 1987.

NEBB Procedural Standards for Certified Testing of Clean Rooms, 1988 Edition.

NFPA Standard No. 45, Fire Protection for Laboratories Using Chemicals, National Fire Protection Association, 1981.

Proceedings of the National Cancer Institute Symposium on Design of Biomedical Research Facilities; Cancer Research Safety Monograph Series, Volume 4.

Selecting a Biological Safety Cabinet, U.S. Department of Health, Education, and Welfare, National Institutes of Health, Division of Safety, Bethesda, MD.

Index